Current Developments in Mathematics 1999

Edited by

David Jerison
Barry Mazur
Wilfried Schmid
Isadore Singer
Daniel Stroock
S.-T. Yau

International Press
www.intlpress.com

Current Developments in Mathematics, 1999

ISBN 978-1-57146-148-3

Paperback reissue 2010. Previously published in 1999 under ISBN 1-57146-043-8 (clothbound).

Typeset using the LaTeX system.

CONTENTS

Preface

These are the proceedings of the joint seminar by M.I.T. and Harvard on the current developments in mathematics for the year 1999. Established in 1995, this seminar will be continued each year.

The organizing committee for the seminar is made up of representatives from the mathematics departments of the two institutions: B. Mazur, W. Schmid, and S.-T. Yau from Harvard; and David Jerison, Isadore Singer, and Dan Stroock from M.I.T.

We would like to thank each of the contributors and to recognise some of the institutions without whose participation the seminar would not have been possible including the departments of amthematics at Harvard and M.I.T. for their financial aid.

Editors

Recent Proofs of the Riemannian Penrose Conjecture

Hubert L. Bray and Richard M. Schoen

In these notes we summarize two recent proofs, by Huisken and Ilmanen [**19**] and by the first author [**6**], of the Riemannian Penrose conjecture in general relativity. The conjecture, which is a statement about the mass of black holes in a three dimensional space-like slice of a spacetime, was first proposed by Penrose [**27**] in 1973 as a test of the cosmic censor conjecture.

In the seventies, Geroch [**11**] and Jang and Wald [**26**] noticed that the Hawking mass of a surface in a 3-manifold was nondecreasing if the surface was flowed out at a rate equal to the inverse of its mean curvature at each point. In addition, they observed that this would be a proof of the Riemannian Penrose conjecture except for the fact that singularites in the flow could develop if the mean curvature ever became zero, preventing the flow from continuing. In 1997, Huisken and Ilmanen found a very nice way to generalize inverse mean curvature flow in such a way that their generalized inverse mean curvature flow always exists, thereby proving the Riemannian Penrose conjecture for a single black hole. (The Huisken-Ilmanen proof allows for multiple black holes, but the bound on the mass is in terms of the area of the largest one.) Then in 1998, using a completely different technique which relies on the positive mass theorem and conformal deformations of metrics, the first author found a proof of the Riemannian Penrose conjecture which works for any number of black holes.

In section 1 we define the important terms and give the statements of the Riemannian positive mass theorem and the Riemannian Penrose conjecture. In section 2, we summarize the first author's proof of the Riemannian Penrose conjecture and provide many of the details. In section 3, we summarize the Huisken-Ilmanen proof, and in section 4 we define the new quasi-local mass quantities which come naturally out of the two approaches.

1. Introduction

General relativity is a theory of gravity which asserts that matter causes four dimensional space-time to be curved, and that our perception of gravity is a consequence of this curvature. Let $(N^4, \bar{g})$ be the space-time manifold with metric $\bar{g}$ of signature $(- + ++)$. Then the central formula of general relativity is Einstein's equation,

(1) $$G = 8\pi T,$$

where T is the stress-energy tensor, $G = Ric(\bar{g}) - \frac{1}{2}R(\bar{g})\cdot\bar{g}$ is the Einstein curvature tensor, $Ric(\bar{g})$ is the Ricci curvature tensor, and $R(\bar{g})$ is the scalar curvature of $\bar{g}$. The beauty of general relativity is that this simple formula explains gravity more accurately than Newtonian physics and is entirely consistent with large scale observations.

However, the nature of the behavior of matter in general relativity is still not well understood. It is not even well understood how to define how much energy and momentum exist in a given region, except in special cases. There does exist a well defined notion of local energy and momentum density which is simply given by the stress-energy tensor which, by equation 1, can be computed in terms of the curvature of N^4. Also, if we assume that the matter of the space-time manifold N^4 is concentrated in some central region of the universe, then N^4 becomes flatter as we get farther away from this central region. If the curvature of N^4 decays quickly enough, then N^4 is said to be asymptotically flat (definition 13 in section 2.4), and with these assumptions it is then possible to define the total mass of the space-time N^4. Interestingly enough, though, the definition of local energy-momentum density, which involves curvature terms of N^4, bears no obvious resemblance to the definition of the total mass of N^4, which is a parameter related to how fast the metric becomes flat at infinity.

The Penrose conjecture [**27**], [**19**], [**6**] and the positive mass theorem ([**31**], [**32**], [**33**], [**34**], [**37**]) can both be thought of as basic attempts at understanding the relationship between the local energy density of a space-time N^4 and the total mass of N^4. In physical terms, the positive mass theorem states that an isolated gravitational system with nonnegative local energy density must have nonnegative total energy. The idea is that nonnegative energy densities must "add up" to something nonnegative. The Penrose conjecture, on the other hand, states that if an isolated gravitational system with nonnegative local energy density contains black holes contributing a mass m, then the total energy of the system must be at least m.

1.1. Statement of the Riemannian Penrose Conjecture.

Important cases of the positive mass theorem and the Penrose conjecture can be translated into statements about complete, asymptotically flat Riemannian 3-manifolds (M^3, g) with nonnegative scalar curvature. If we consider (M^3, g, h) as a space-like hypersurface of $(N^4, \bar{g})$ with metric g_{ij} and second fundamental form h_{ij} in N^4, then equation 1 implies that

$$\mu = G_0^0 = \frac{1}{16\pi}[R - \sum_{i,j} h^{ij}h_{ij} + (\sum_i h_i^i)^2], \tag{2}$$

$$J^i = G_0^i = \frac{1}{8\pi}\sum_j \nabla_j[h^{ij} - (\sum_k h_k^k)g^{ij}], \tag{3}$$

where R is the scalar curvature of the metric g, μ is the local energy density, and J^i is the local current density. The assumption of nonnegative energy density everywhere in N^4 implies that we must have

$$\mu \geq \left(\sum_i J^i J_i\right)^{\frac{1}{2}} \tag{4}$$

at all points on M^3. Equations 2, 3, and 4 are called the constraint equations for (M^3, g, h) in $(N^4, \bar{g})$. Thus we see that if we restrict our attention to 3-manifolds which have zero second fundamental form h in N^3, the constraint equations are equivalent to the condition that the Riemannian manifold (M^3, g) has nonnegative scalar curvature everywhere.

A Riemannian manifold (M^3, g) is said to be asymptotically flat if, outside a compact set, it is the disjoint union of one or more regions (called ends) diffeomorphic to $(\mathbf{R}^3 \backslash B_1(0), \delta)$, where the metric g in each of these $\mathbf{R}^3$ coordinate charts approaches the standard metric δ on $\mathbf{R}^3$ at infinity (with certain asymptotic decay conditions - see definition 13, [**33**], [**2**]). The positive mass theorem and the Penrose conjecture are both statements which refer to a particular **chosen end** of (M^3, g). The total mass of (M^3, g) is then a parameter related to how fast this chosen end of (M^3, g) becomes flat at infinity. The most general definition of the total mass (also called the ADM mass) is given in equation 59 in section 2.4. In addition, we give an alternative definition of total mass in the next subsection.

The positive mass theorem was first proved by Schoen and Yau [**31**] in 1979 using minimal surfaces. The Riemannian positive mass theorem is a special case of the positive mass theorem which comes from considering the space-like hypersurfaces which have zero second fundamental form in the spacetime.

The Riemannian Positive Mass Theorem
Let (M^3, g) be a complete, smooth, asymptotically flat 3-manifold with nonnegative scalar curvature and total mass m. Then

$$m \geq 0, \tag{5}$$

with equality if and only if (M^3, g) is isometric to $\mathbf{R}^3$ with the standard flat metric.

Apparent horizons of black holes in N^4 correspond to outermost minimal surfaces of M^3 if we assume M^3 has zero second fundamental form in N^4. A minimal surface is a surface which has zero mean curvature (and hence is a critical point for the area functional). An outermost minimal surface is a minimal surface which is not contained entirely inside another minimal surface. Again, there is a chosen end of M^3, and "contained entirely inside" is defined with respect to this end. Interestingly, it follows from a stability argument [**16**] that outermost minimal surfaces are always spheres. There could be more than one outermost sphere, with each minimal sphere corresponding to a different black hole, and we note that outermost minimal spheres never intersect.

As an example, consider the Schwarzschild manifolds $(\mathbf{R}^3 \backslash \{0\}, s)$ where $s_{ij} = (1 + m/2r)^4\, \delta_{ij}$ and m is a positive constant and equals the total mass of the manifold. This manifold has zero scalar curvature everywhere, is spherically symmetric, and it can be checked that it has an outermost minimal sphere at $r = m/2$.

We define the horizon of (M^3, g) to be the union of all of the outermost minimal spheres in M^3, so that the horizon of a manifold can have multiple connected components. We note that it is usually more common to call each outermost minimal sphere a horizon, so that their union is referred to as "horizons", but it turns out to be mathematically more convenient for our purposes to refer to the union of all of the outermost minimal spheres as one object, which we will call the horizon of (M^3, g).

There is a very convincing (but not rigorous) physical motivation to define the mass that a collection of black holes contributes to be $\sqrt{\frac{A}{16\pi}}$, where A is the total surface area of the horizon of (M^3, g). Then the physical statement that a system with nonnegative energy density containing black holes contributing a mass m must have total mass at least m can be translated into the following geometric statement [**27**], [**26**].

The Riemannian Penrose Conjecture
Let (M^3, g) be a complete, smooth, asymptotically flat 3-manifold with nonnegative scalar curvature and total mass m whose outermost minimal spheres have total surface area A. Then

$$m \geq \sqrt{\frac{A}{16\pi}}, \tag{6}$$

with equality if and only if (M^3, g) is isometric to the Schwarzschild metric $(\mathbf{R}^3 \backslash \{0\}, s)$ of mass m outside their respective horizons.

The first breakthrough on the Riemannian Penrose conjecture was made by Huisken and Ilmanen who proved the above theorem in the case that the horizon of (M^3, g) has only one component [**19**]. Specifically, they proved that $m \geq \sqrt{\frac{A_{max}}{16\pi}}$, where A_{max} is the area of the largest component of the horizon of (M^3, g). Their proof is as interesting as the result itself. In the seventies, Geroch [**11**] observed that in a manifold with nonnegative scalar curvature, the Hawking mass of a sphere (but not surfaces with multiple components) was monotone increasing under a $1/H$ flow, where H is the mean curvature of the sphere. Jang and Wald [**26**] proposed using this to attack the Riemannian Penrose conjecture by flowing the horizon of the manifold out to infinity. However, it is not hard to concoct situations in which the $1/H$ flow of a sphere develops singularities, preventing the idea from working much of the time. Huisken and Ilmanen's approach then was to define a generalized $1/H$ flow which sometimes "jumps" in order to prevent singularities from developing. Their techniques are described in section 3.

In section 2, we sketch the proof of the Riemannian Penrose conjecture due to the first author. The overview of the proof of this result is given in the introduction to section 2. The basic idea of the approach to the problem is to flow the original metric continuously to a Schwarzschild metric (outside the horizon). The particular flow we define has the important property that the area of the horizon stays constant while the total mass of the manifold is non-increasing. Then since the Schwarzschild metric gives equality in the Penrose inequality, the inequality follows for the original metric.

Other contributions have also been made by Herzlich [**18**] using the Dirac operator which Witten [**37**] used to prove the positive mass theorem, by Gibbons [**12**] in the special case of collapsing shells, by Tod [**36**], by Bartnik [**4**] for quasi-spherical metrics, and by the first author [**5**] using isoperimetric surfaces.

1.2. Definitions and Setup. Without loss of generality, we will be able to assume that an asymptotically flat metric (see definition 13) has an even nicer form at infinity in each end because of the following lemma and definition.

LEMMA 1. **(Schoen, Yau [33])** *Let (M^3, g) be any asymptotically flat metric with nonnegative scalar curvature. Then given any $\epsilon > 0$, there exists a metric g_0*

with nonnegative scalar curvature which is harmonically flat at infinity (defined in the next definition) such that

$$1 - \epsilon \leq \frac{g_0(\vec{v},\vec{v})}{g(\vec{v},\vec{v})} \leq 1 + \epsilon \tag{7}$$

for all nonzero vectors $\vec{v}$ in the tangent space at every point in M and

$$|\bar{m}_k - m_k| \leq \epsilon \tag{8}$$

where $\bar{m}_k$ and m_k are respectively the total masses of (M^3, g_0) and (M^3, g) in the kth end.

Notice that because of equation 7, the percentage difference in areas as well as lengths between the two metrics is arbitrarily small. Hence, since the mass changes arbitrarily little also and since inequality 6 is a closed condition, it follows that the Riemannian Penrose inequality for asymptotically flat manifolds follows from proving the inequality for manifolds which are harmonically flat at infinity.

DEFINITION 1. *A Riemannian manifold is defined to be* **harmonically flat at infinity** *if, outside a compact set, it is the disjoint union of regions (which we will again call ends) with zero scalar curvature which are conformal to $(\mathbf{R}^3 \backslash B_1(0), \delta)$ with the conformal factor approaching a positive constant at infinity in each region.*

Now it is fairly easy to define the total mass of an end of a manifold (M^3, g_0) which is harmonically flat at infinity. Define g_{flat} to be a smooth metric on M^3 conformal to g_0 such that in each end of M^3 in the above definition (M^3, g_{flat}) is isometric to $(\mathbf{R}^3 \backslash B_1(0), \delta)$. Define $\mathcal{U}_0(x)$ such that

$$g_0 = \mathcal{U}_0(x)^4 g_{flat}. \tag{9}$$

Since (M^3, g_0) has zero scalar curvature in each end, $(\mathbf{R}^3 \backslash B_1(0), \mathcal{U}_0(x)^4 \delta)$ must have zero scalar curvature. This implies that $\mathcal{U}_0(x)$ is harmonic in $(\mathbf{R}^3 \backslash B_1(0), \delta)$ (see equation 95 in appendix E). Since $\mathcal{U}_0(x)$ is a harmonic function going to a constant at infinity, we may expand it in terms of spherical harmonics to get

$$\mathcal{U}_0(x) = a + \frac{b}{|x|} + \mathcal{O}\left(\frac{1}{|x|^2}\right), \tag{10}$$

where a and b are constants.

DEFINITION 2. *The* **total mass** *(of an end) of a Riemannian 3-manifold which is harmonically flat at infinity is defined to be $2ab$ in the above equation.*

While the constants a and b scale depending on how we represent (M^3, g_0) as the disjoint union of a compact set and ends in definition 1, it can be checked that $2ab$ does not. Furthermore, this definition agrees with the standard definition of the total mass of an asymptotically flat manifold (defined in equation 59) in the case that the manifold is harmonically flat at infinity. In section 2 we will choose to work with definition 2 because it is more convenient for the calculations we will be doing.

Now we turn our attention to the definition and properties of horizons. For convenience, we modify the topology of M^3 by compactifying all of the ends of M^3 except for the chosen end by adding the points $\{\infty_k\}$.

DEFINITION 3. *Define $\mathcal{S}$ to be the collection of surfaces which are smooth compact boundaries of open sets in M^3 containing the points $\{\infty_k\}$.*

All of the surfaces that we will be dealing with in section 2 will be in $\mathcal{S}$. Also, we see that all of the surfaces in $\mathcal{S}$ divide M^3 into two regions, an inside (the open set) and an outside (the complement of the open set). Thus, the notion of one surface in $\mathcal{S}$ (entirely) enclosing another surface in $\mathcal{S}$ is well defined.

DEFINITION 4. *A* **horizon** *of (M^3, g) is any zero mean curvature surface in $\mathcal{S}$.*

A horizon may have multiple components. Furthermore, by minimizing area over surfaces in $\mathcal{S}$, a horizon is guaranteed to exist when M^3 has more than one end.

DEFINITION 5. *A horizon is defined to be* **outermost** *if it is not enclosed by another horizon.*

We note that when at least one horizon exists, there is always a unique outermost horizon, with respect to the chosen end.

DEFINITION 6. *A surface $\Sigma \in \mathcal{S}$ is definedto be* **outer-minimizing** *if every other surface $\tilde{\Sigma} \in \mathcal{S}$ which encloses it has $|\tilde{\Sigma}| \geq |\Sigma|$.*

An outer-minimizing surface must have nonnegative mean curvature since otherwise the first variation formula would imply that an outward variation would yield a surface with less area. Also, in the case that Σ is a horizon, it is an outer-minimizing horizon if and only if it is not enclosed by a horizon with less area. We also note that, as with outermost horizons, it follows from a second variation argument that each component of an outer-minimizing horizon must be a sphere **[30]**. We will not use outermost horizons in section 2, but simply point out that outermost horizons are always outer-minimizing. Hence, the following theorem, which we will prove in section 2, is a slight generalization of the Riemannian Penrose conjecture.

THEOREM 1. *Let (M^3, g) be a complete, smooth, asymptotically flat 3-manifold with nonnegative scalar curvature, total mass m, and an outer-minimizing horizon (with one or more components) of total area A. Then*

$$m \geq \sqrt{\frac{A}{16\pi}} \tag{11}$$

with equality if and only if (M^3, g) is isometric to a Schwarzschild manifold outside their respective outermost horizons.

Besides the flat metric on $\mathbf{R}^3$, the Schwarzschild manifolds are the only other complete spherically symmetric 3-manifolds with zero scalar curvature, and as previously mentioned can be described explicitly as $(\mathbf{R}^3 \backslash \{0\}, s)$ where

$$s_{ij} = \left(1 + \frac{m}{2r}\right)^4 \delta_{ij},$$

r is the distance from the origin in $\mathbf{R}^3$, and m is a positive constant and equals the total mass of the manifold. Then since the Schwarzschild manifolds have a single minimal sphere which is the coordinate sphere of radius $m/2$, we can verify they give equality in the above theorem.

2. Proof of the Riemannian Penrose Conjecture Using the Positive Mass Theorem

In this section we give the overview of the proof of theorem 1 which is a slight generalization of the Riemannian Penrose conjecture [6]. As discussed in the previous section, without loss of generality for proving the Riemannian Penrose inequality we may make the following convenient, but still quite general, assumption.

Assumptions: For the rest of this section, we will assume that (M^3, g_0) has nonnegative scalar curvature and is harmonically flat at infinity, unless otherwise stated.

We will then generalize our results to the asymptotically flat case and handle the case of equality of theorem 1 in section 2.4.

We define a continuous family of conformal metrics $\{g_t\}$ on M^3, where

$$g_t = u_t(x)^4 g_0 \tag{12}$$

and $u_0(x) \equiv 1$. Given the metric g_t, define

$$\Sigma(t) = \text{the outermost minimal area enclosure of } \Sigma_0 \text{ in } (M^3, g_t) \tag{13}$$

where Σ_0 is the original outer-minimizing horizon in (M^3, g_0) and we stay inside the collection of surfaces $\mathcal{S}$ defined in the previous section. In the cases in which we are interested, $\Sigma(t)$ will not touch Σ_0, from which it follows that $\Sigma(t)$ is actually an outer-minimizing horizon of (M^3, g_t). Then given the horizon $\Sigma(t)$, define $v_t(x)$ such that

$$\begin{cases} \Delta_{g_0} v_t(x) & \equiv & 0 & \text{outside } \Sigma(t) \\ v_t(x) & = & 0 & \text{on } \Sigma(t) \\ \lim_{x\to\infty} v_t(x) & = & -e^{-t} & \end{cases} \tag{14}$$

and $v_t(x) \equiv 0$ inside $\Sigma(t)$. Finally, given $v_t(x)$, define

$$u_t(x) = 1 + \int_0^t v_s(x)ds \tag{15}$$

so that $u_t(x)$ is continuous in t and has $u_0(x) \equiv 1$.

THEOREM 2. *Taken together, equations 12, 13, 14, and 15 define a first order o.d.e. in t for $u_t(x)$ which has a solution which is Lipschitz in the t variable, C^1 in the x variable everywhere, and smooth in the x variable outside $\Sigma(t)$. Furthermore, $\Sigma(t)$ is a smooth, outer-minimizing horizon in (M^3, g_t) for all $t \geq 0$, and $\Sigma(t_2)$ encloses but does not touch $\Sigma(t_1)$ for all $t_2 > t_1 \geq 0$.*

Proof. See [6].

Since $v_t(x)$ is a superharmonic function in (M^3, g_0), it follows that $u_t(x)$ is superharmonic as well, and from equation 15 we see that $\lim_{x\to\infty} u_t(x) = e^{-t}$ and consequently that $u_t(x) > 0$ for all t. Then since

$$R(g_t) = u_t(x)^{-5}(-8\Delta_g + R(g))u_t(x) \tag{16}$$

it follows that (M^3, g_t) is an asymptotically flat manifold with nonnegative scalar curvature.

Even so, it still may not seem like g_t is particularly naturally defined since the rate of change of g_t appears to depend on t and the original metric g_0 in equation 14. We would prefer a flow where the rate of change of g_t is only a function of g_t

(and M^3 and Σ_0 perhaps), and interestingly enough this actually does turn out to be the case. In appendix E we prove this very important fact and provide a very natural motivation for defining this conformal flow of metrics.

DEFINITION 7. *The function $A(t)$ is defined to be the total area of the horizon $\Sigma(t)$ in (M^3, g_t).*

DEFINITION 8. *The function $m(t)$ is defined to be the total mass of (M^3, g_t) in the chosen end.*

The next theorem is the key property of the conformal flow of metrics.

THEOREM 3. *The function $A(t)$ is constant in t and $m(t)$ is non-increasing in t, for all $t \geq 0$.*

Proof. The fact that $A'(t) = 0$ follows from the fact that to first order the metric is not changing on $\Sigma(t)$ (since $v_t(x) = 0$ there) and from the fact that to first order the area of $\Sigma(t)$ does not change as it moves outward since $\Sigma(t)$ has zero mean curvature in (M^3, g_t). Hence, $A(t)$ is constant by construction really. We refer the reader to **[6]** for the rigorous argument. The interesting part of theorem 3 is proving that $m'(t) \leq 0$. Curiously, this follows from a nice trick using the Riemannian positive mass theorem and is explained in detail in sections 2.2 and 2.3. □

Another important aspect of this conformal flow of the metric (which we do not have room to prove in these notes) is that outside the horizon $\Sigma(t)$, the manifold (M^3, g_t) becomes more and more spherically symmetric and "approaches" a Schwarzschild manifold $(\mathbf{R}^3 \backslash \{0\}, s)$ in the limit as t goes to ∞. More precisely,

THEOREM 4. *For sufficiently large t, there exists a diffeomorphism ϕ_t between (M^3, g_t) outside the horizon $\Sigma(t)$ and a fixed Schwarzschild manifold $(\mathbf{R}^3 \backslash \{0\}, s)$ outside its horizon. Furthermore, for all $\epsilon > 0$, there exists a T such that for all $t > T$, the metrics g_t and $\phi_t^*(s)$ (when determining the lengths of unit vectors of (M^3, g_t)) are within ϵ of each other and the total masses of the two manifolds are within ϵ of each other. Hence,*

$$\lim_{t \to \infty} \frac{m(t)}{\sqrt{A(t)}} = \sqrt{\frac{1}{16\pi}}. \tag{17}$$

Proof. See **[6]**.

Inequality 11 of theorem 1 then follows from theorems 2, 3 and 4, for harmonically flat manifolds. In addition, in section 2.4 we will see that the case of equality in theorem 1 follows from the fact that $m'(0) = 0$ if and only if (M^3, g_0) is isometric to a Schwarzschild manifold outside their respective outermost horizons. We will also generalize our results to the asymptotically flat case in section 2.4.

2.1. Some Qualitative Features of the Conformal Flow of Metrics.

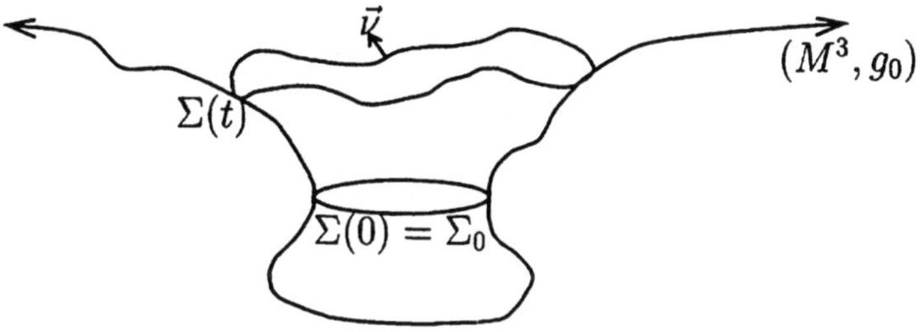

The diagrams above and below are meant to help illustrate some of the properties of the conformal flow of the metric. The above picture is the original metric which has an outer-minimizing horizon Σ_0. As t increases, $\Sigma(t)$ moves outwards, but never inwards. In the diagram below, we can observe one of the consequences of the fact that $A(t) = A_0$ is constant in t. Since the metric is not changing inside $\Sigma(t)$, all of the horizons $\Sigma(s)$, $0 \leq s \leq t$ have area A_0 in (M^3, g_t). Hence, inside $\Sigma(t)$, the manifold (M^3, g_t) becomes cylinder-like in the sense that it is laminated by all of the previous horizons which all have the same area A_0 with respect to the metric g_t.

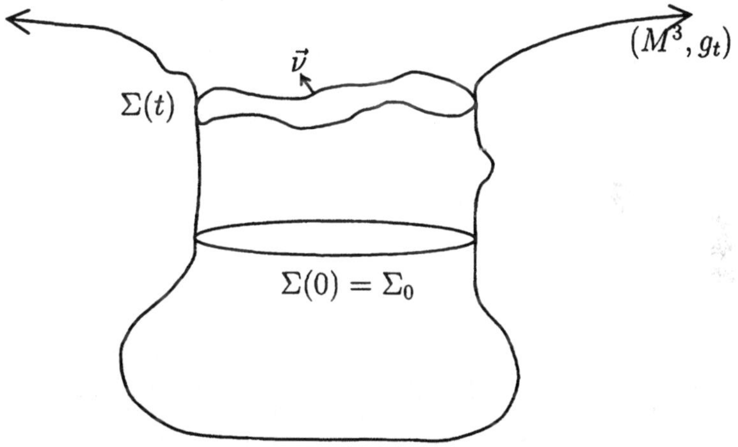

Now let us suppose that the original horizon Σ_0 of (M^3, g) had two components, for example. Then each of the components of the horizon will move outwards as t increases, and at some point before they touch they will suddenly jump outwards to form a horizon with a single component enclosing the previous horizons with two components. Even horizons with only one component will sometimes jump outwards, and it is interesting that this phenomenon of surfaces jumping is also found in the Huisken-Ilmanen approach to the Penrose conjecture using their generalized $1/H$ flow.

We end this subsection with two definitions which will be needed later.

DEFINITION 9. *Define*

$$\Sigma^+(t) = \lim_{s\to t^+} \Sigma(s), \qquad \Sigma^-(t) = \lim_{s\to t^-} \Sigma(s), \tag{18}$$

We note that since $\Sigma(t_2)$ encloses $\Sigma(t_1)$ for all $t_2 > t_1 \geq 0$ by theorem 2, the above limits always exist and are unique. Then we can define the jump times for the horizons $\Sigma(t)$ very naturally.

DEFINITION 10. *Define the jump times J to be the set of all $t \geq 0$ with $\Sigma^+(t) \neq \Sigma^-(t)$, where we define $\Sigma^-(0) = \Sigma_0$.*

2.2. Green's Functions on Asy. Flat 3-Manifolds with Nonnegative Scalar Curvature and the Riemannian Positive Mass Theorem. In this subsection we will prove a few theorems about certain Green's functions on asymptotically flat 3-manifolds with nonnegative scalar curvature which will be needed in the next subsection to prove that $m(t)$ is nonincreasing in t. However, the theorems in this subsection, which follow from and generalize the Riemannian positive mass theorem, are also of independent interest.

The results of this subsection are closely related to the beautiful ideas used by Bunting and Masood-ul-Alam in [8] to prove the non-existence of multiple black holes in asymptotically flat, static, vacuum space-times. Hence, while theorems 5 and 6 do not appear in their paper, these two theorems follow from a natural extension of their techniques.

DEFINITION 11. *Given a complete, asymptotically flat manifold $(M^3, \bar{g})$ with multiple asymptotically flat ends (with one chosen end), define*

$$\mathcal{E}(\bar{g}) = \inf_{\phi} \left\{ \frac{1}{2\pi} \int_{(M^3,\bar{g})} |\nabla\phi|^2 \, dV \right\} \tag{19}$$

where the infimum is taken over all smooth $\phi(x)$ which go to one in the chosen end and zero in the other ends.

Without loss of generality, we may assume that $(M^3, \bar{g})$ is actually harmonically flat at infinity (as defined in section 1.2). Then since such a modification can be done so as to change the metric uniformly pointwise as small as one likes (by lemma 1), it follows that $\mathcal{E}(\bar{g})$ changes as small as one likes as well. We remind the reader that the total mass of $(M^3, \bar{g})$ also changes arbitrarily little with such a deformation.

From standard theory it follows that the infimum in the above definition is achieved by the Green's function $\phi(x)$ which satisifies

$$\begin{cases} \lim_{x\to\infty_0} \phi(x) &= 1 \\ \Delta\phi &= 0 \\ \lim_{x\to\infty_k} \phi(x) &= 0 \text{ for all } k \neq 0 \end{cases} \tag{20}$$

where ∞_k are the points at infinity of the various asymptotically flat ends and ∞_0 is infinity in the chosen end. Define the level sets of $\phi(x)$ to be

$$\Sigma_l = \{x \mid \phi(x) = l\} \tag{21}$$

for $0 < l < 1$. Then it follows from Sard's theorem and the smoothness of $\phi(x)$ that $\Sigma(l)$ is a smooth surface for almost every l. Then by the co-area formula it follows that

$$\mathcal{E}(\bar{g}) = \frac{1}{2\pi}\int_0^1 dl \int_{\Sigma_l} |\nabla\phi|\, dA = \frac{1}{2\pi}\int_0^1 dl \int_{\Sigma_l} \frac{d\phi}{d\eta}\, dA \tag{22}$$

since $\nabla\phi$ is orthogonal to the unit normal vector η of Σ_l. But by the divergence theorem (and since $\phi(x)$ is harmonic), $\int_\Sigma \frac{d\phi}{d\eta}\, dA$ is constant for all homologous Σ. Hence,

$$\mathcal{E}(\bar{g}) = \frac{1}{2\pi}\int_\Sigma \frac{d\phi}{d\eta}\, dA \tag{23}$$

where Σ is any surface in M^3 in $\mathcal{S}$ (which is the set of smooth, compact boundaries of open regions which contains the points at infinity $\{\infty_k\}$ in all of the ends except the chosen one). Then since $(M^3, \bar{g})$ is harmonically flat at infinity, we know that in the chosen end $\phi(x) = 1 - c/|x| + \mathcal{O}(1/|x|^2)$, so that from equation 23 it follows that

$$\phi(x) = 1 - \frac{\mathcal{E}(\bar{g})}{2|x|} + \mathcal{O}\left(\frac{1}{|x|^2}\right) \tag{24}$$

by letting Σ in equation 23 be a large sphere in the chosen end.

THEOREM 5. *Let $(M^3, \bar{g})$ be a complete, smooth, asymptotically flat 3-manifold with nonnegative scalar curvature which has multiple asymptotically flat ends and total mass $\bar{m}$ in the chosen end. Then*

$$\bar{m} \geq \mathcal{E}(\bar{g}) \tag{25}$$

with equality if and only if $(M^3, \bar{g})$ has zero scalar curvature and is conformal to $(\mathbf{R}^3, \delta)$ minus a finite number of points.

Proof. Again, without loss of generality we will assume that $(M^3, \bar{g})$ is harmonically flat at infinity, which by definition means that all of the ends are harmonically flat at infinity. In the picture below, $(M^3, \bar{g})$ has three ends, the chosen end (at the top of the picture) and two other ends.

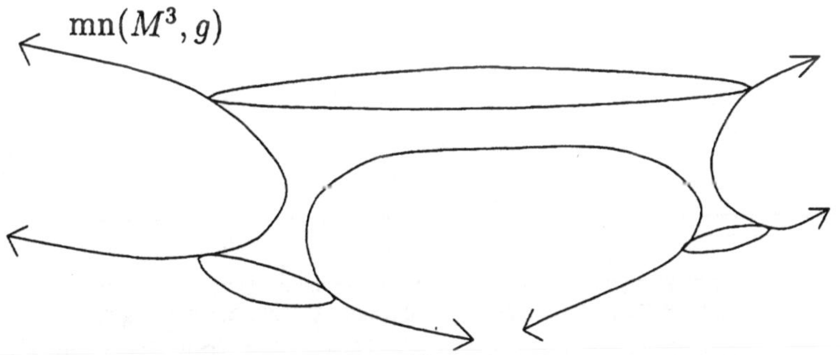

Then we consider the metric $(M^3, \tilde{g})$, with $\tilde{g} = \phi(x)^4 \bar{g}$, drawn below. Since $\phi(x)$ goes to zero (and is bounded above by $C/|x|$) in all of the harmonically flat ends other than the chosen one, the metric $\tilde{g} = \phi(x)^4 \bar{g}$ in each end is conformal to a punctured ball with the conformal factor being a bounded harmonic function to the fourth power in the punctured ball. Hence, by the removable singularity theorem, this harmonic function can be extended to the whole ball, which proves that the metric $\tilde{g}$ can be extended smoothly over all of the points at infinity in the compactified ends.

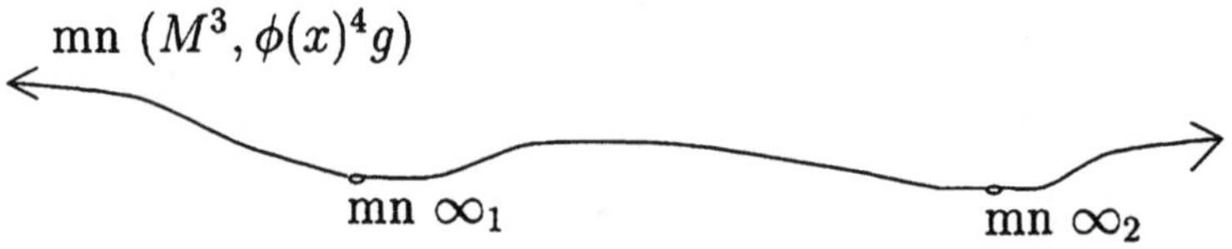

Furthermore, $(M^3, \tilde{g})$ has nonnegative scalar curvature since $(M^3, \bar{g})$ has nonnegative scalar curvature and $\phi(x)$ is harmonic with respect to $\bar{g}$ (see equation 95 in appendix E). Moreover, $(M^3 \cup \{\infty_k\}, \tilde{g})$ has nonnegative scalar curvature too since in a neighborhood of each ∞_k the manifold is conformal to a ball, with the conformal factor being a positive harmonic function to the fourth power. Hence, since $(M^3 \cup \{\infty_k\}, \tilde{g})$ is a complete 3-manifold with nonnegative scalar curvature with a single harmonically flat end, we may apply the Riemannian positive mass theorem to this manifold to conclude that the total mass of this manifold, which we will call $\tilde{m}$, is nonnegative.

Now we will compute $\tilde{m}$ in terms of $\bar{m}$ and $\mathcal{E}(\bar{g})$. Since $\bar{g}$ is harmonically flat at infinity, we know that by definition 2 we have $\bar{g} = \bar{\mathcal{U}}(x)^4 \bar{g}_{flat}$, where $(M^3, \bar{g}_{flat})$ is isometric to $(\mathbf{R}^3 \backslash B_r(0), \delta)$ in the harmonically flat end of $(M^3, \bar{g})$,

$$\bar{\mathcal{U}}(x) = 1 + \frac{\bar{m}}{2|x|} + \mathcal{O}\left(\frac{1}{|x|^2}\right), \tag{26}$$

and the scale of the harmonically flat coordinate chart $\mathbf{R}^3 \backslash B_r(0)$ has been chosen so that $\bar{\mathcal{U}}(x)$ goes to one at infinity in the chosen end. Furthermore, since $\tilde{g} = \phi(x)^4 \bar{g}$ is also harmonically flat at infinity (which follows from equation 94 in appendix E), we have $\tilde{g} = \tilde{\mathcal{U}}(x)^4 \bar{g}_{flat} = \phi(x)^4 \bar{\mathcal{U}}(x)^4 \bar{g}_{flat}$ where

$$\tilde{\mathcal{U}}(x) = 1 + \frac{\tilde{m}}{2|x|} + \mathcal{O}\left(\frac{1}{|x|^2}\right). \tag{27}$$

Then comparing equations 24, 26, and 27 with $\tilde{\mathcal{U}}(x) = \bar{\mathcal{U}}(x)\phi(x)$ yields

$$\tilde{m} = \bar{m} - \mathcal{E}(\bar{g}) \geq 0 \tag{28}$$

by the Riemannian positive mass theorem, which proves inequality 25 for harmonically flat manifolds. Then since asymptotically flat manifolds can be arbitrarily well approximated by harmonically flat manifolds by lemma 1, inequality 25 follows for asymptotically flat manifolds as well.

To prove the case of equality, we require a generalization of the case of equality of the positive mass theorem given in [7] as theorem 5.3. In that paper, we say that a singular manifold has *generalized nonnegative scalar curvature* if it is the limit (in the sense given in [7]) of smooth manifolds with nonnegative scalar curvature.

Note that if $(M^3, \bar{g})$ is only asymptotically flat and not harmonically flat, then we have not shown that $(M^3, \tilde{g})$ can be extended smoothly over the missing points $\{\infty_k\}$. However, as a possibly singular manifold, it does have generalized nonnegative scalar curvature since it is the limit of smooth manifolds with nonnegative scalar curvature (since $(M^3, \bar{g})$ can be arbitrarily well approximated by harmonically flat manifolds).

Then theorem 5.3 in [7] states that if $(M^3, \tilde{g})$ has generalized nonnegative scalar curvature, zero mass, and positive isoperimetric constant, then $(M^3, \tilde{g})$ is flat (outside the singular set). Hence, it follows that if we have equality in inequality 25, then $\tilde{g}$ is flat. Hence, $(M^3, \tilde{g})$ is isometric to $(\mathbf{R}^3, \delta)$ minus a finite number of points, and since the harmonic conformal factor $\phi(x)$ preserves the sign of the scalar curvature by equation 95, the case of equality of the theorem follows. □

DEFINITION 12. *Given a complete, asymptotically flat manifold (M^3, g) with horizon $\Sigma \in \mathcal{S}$ (defined in section 1.2), define*

$$\mathcal{E}(\Sigma, g) = \inf_{\varphi} \left\{ \frac{1}{2\pi} \int_{M^3} |\nabla\varphi|^2 \, dV \right\} \tag{29}$$

where the infimum is taken over all smooth $\varphi(x)$ which go to one at infinity and equal zero on the horizon Σ (and are zero inside Σ). (By definition of $\mathcal{S}$, all of the ends other than the chosen end are contained inside Σ.)

The infimum in the above definition is achieved by the Green's function $\varphi(x)$ which satisifies

$$\begin{cases} \lim_{x\to\infty_0} \varphi(x) &= 1 \\ \Delta\varphi &= 0 \\ \varphi(x) &= 0 \text{ on } \Sigma \end{cases} \tag{30}$$

and as before,

$$\varphi(x) = 1 - \frac{\mathcal{E}(\Sigma, g)}{2|x|} + \mathcal{O}\left(\frac{1}{|x|^2}\right) \tag{31}$$

in the chosen end.

THEOREM 6. *Let (M^3, g) be a complete, smooth, asymptotically flat 3-manifold with nonnegative scalar curvature with a horizon $\Sigma \in \mathcal{S}$ and total mass m (in the chosen end). Then*

$$m \geq \frac{1}{2}\mathcal{E}(\Sigma, g) \tag{32}$$

with equality if and only if (M^3, g) is a Schwarzschild manifold outside the horizon Σ.

Proof. Let M^3_Σ be the closed region of M^3 which is outside (or on) Σ. Since $\Sigma \in \mathcal{S}$, (M^3_Σ, g) has only one end, and we recall that Σ could have multiple components. For example, in the picture below Σ has two components.

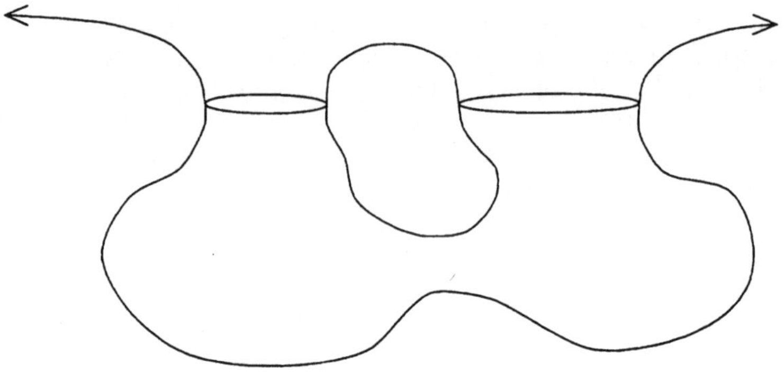

Then the basic idea is to reflect (M^3_Σ, g) through Σ to get a manifold $(\bar{M}^3_\Sigma, \bar{g})$ with two asymptotically flat ends. Then define $\phi(x)$ on $(\bar{M}^3_\Sigma, \bar{g})$ using equation 20 and $\varphi(x)$ on (M^3_Σ, g) using equation 30. It follows from symmetry that $\phi(x) = \frac{1}{2}$ on Σ, so that

$$\phi(x) = \frac{1}{2}(\varphi(x) + 1) \tag{33}$$

on (M^3_Σ, g). Then

$$\mathcal{E}(\bar{g}) = \frac{1}{2}\mathcal{E}(\Sigma, g) \tag{34}$$

so that theorem 6 follows from theorem 5.

The only technicality is that theorem 5 applies to smooth manifolds with nonnegative scalar curvature, and $(\bar{M}^3_\Sigma, \bar{g})$ is typically not smooth along Σ, which also makes it unclear how to define the scalar curvature there. However, it happens that because Σ has zero mean curvature, these issues can be resolved.

This idea of reflecting a manifold through its horizon is used by Bunting and Masood-ul-Alam in [**8**], and the issue of the smoothness of the reflected manifold appears in their paper as well. However, in their setting they have the simpler case in which the horizon not only has zero mean curvature but also has zero second fundamental form. Hence, the reflected manifold is $C^{1,1}$, which apparently is sufficient for their purposes.

However, in our setting we can not assume that the horizon Σ has zero second fundamental form, so that $(\bar{M}^3_\Sigma, \bar{g})$ is only Lipschitz. To solve this problem, given $\delta > 0$ we will define a smooth manifold $(\tilde{M}^3_{\Sigma,\delta}, \tilde{g}_\delta)$ with nonnegative scalar curvature which, in the limit as δ approaches zero, approaches $(\bar{M}^3_\Sigma, \bar{g})$ (meaning that there exists a diffeomorphism under which the metrics are arbitrarily uniformly close to each other and the total masses are arbitrarily close). Then by definition 11 it follows that $\mathcal{E}(\tilde{g}_\delta)$ is close to $\mathcal{E}(\bar{g})$, from which we will be able to conclude

$$m \approx \tilde{m}_\delta \geq \mathcal{E}(\tilde{g}_\delta) \approx \mathcal{E}(\bar{g}) = \frac{1}{2}\mathcal{E}(\Sigma, g), \tag{35}$$

where $\tilde{m}_\delta$ is the mass of $(\tilde{M}^3_{\Sigma,\delta}, \tilde{g}_\delta)$ and the approximations in the above inequality can be made to be arbitrarily accurate by choosing δ small, thereby proving inequality 32.

The first step is to constuct the smooth manifolds

$$(\tilde{M}^3_{\Sigma,\delta}, \tilde{g}_\delta) \approx (M^3_\Sigma, g) \cup (\Sigma \times [0, 2\delta], G) \cup (M^3_\Sigma, g), \tag{36}$$

where idenifications are made along the boundaries of these three manifolds as drawn below. (To be precise, the second (M^3_Σ, g) in the above union is meant to be a copy of the first (M^3_Σ, g) and therefore distinct.) We will define the metric G such that the metric $\bar{g}_\delta$ is smooth, although it will not have nonnegative scalar curvature. Then we will define

$$\tilde{g}_\delta = u_\delta(x)^4 \bar{g}_\delta \tag{37}$$

so that $\tilde{g}_\delta$ is not only smooth but also has nonnegative scalar curvature, and we will show that because of our choice of the metric G, $u_\delta(x)$ approaches one in the limit as δ approaches zero.

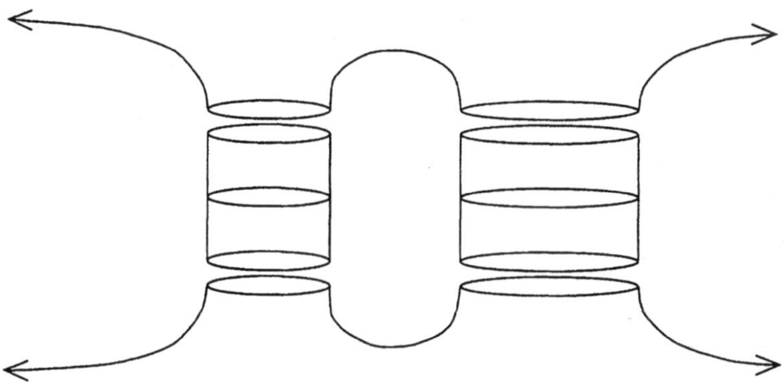

We will use the local coordinates (z,t) to describe points on $\Sigma \times [0, 2\delta]$, where $z = (z_1, z_2) \in$ a local coordinate chart for Σ and $t \in [0, 2\delta]$. Then we define $G(\partial_t, \partial_t) = 1$, $G(\partial_t, \partial_{z_1}) = 0$, and $G(\partial_t, \partial_{z_2}) = 0$. Then it follows that $\Sigma \times t$ is obtained by flowing $\Sigma \times 0$ in the unit normal direction for a time t, and that ∂_t is orthogonal to $\Sigma \times t$. Hence, all that remains to fully define the metric G is to define it smoothly on the tangent planes of $\Sigma \times t$ for $0 \le t \le 2\delta$.

Let $\bar{G}(z,t)$ be the metric G restricted to $\Sigma \times t$. Then

$$\frac{d}{dt}\bar{G}_{ij}(z,t) = 2\bar{G}_{ik}(z,t)h^k_j(z,t) \tag{38}$$

where $h^k_j(z,t)$ is the second fundamental form of $\Sigma \times t$ in $(\Sigma \times [0,2\delta], G)$ with respect to the normal vector ∂_t. Furthermore, since $\Sigma \times 0$ is identified with $\Sigma \in (M^3_\Sigma, g)$, we can extend the coordinates (z,t) for t slightly less than zero into (M^3_Σ, g), thereby giving us smooth initial data for $\bar{G}_{ij}(z,t)$ and $h^k_j(z,t)$ for $-\epsilon < t \le 0$, for some positive ϵ.

Now we extend $h^k_j(z,t)$ smoothly for $0 \le t \le 2\delta$ in such a way that $h^k_j(z,t)$ is an odd function about $t = \delta$, meaning that $h^k_j(z,t) = -h^k_j(z, 2\delta - t)$. Naturally there are many ways to accomplish this smooth extension.

Then we define $\bar{G}_{ij}(z,t)$ to be the smooth solution to the o.d.e. given in equation 38 using the initial data for $\bar{G}_{ij}(z,t)$ at $t = 0$. By the oddness of $h^k_j(z,t)$ about $t = \delta$ it follows that $\bar{G}_{ij}(z,t)$ is symmetric about $t = \delta$, that is, $\bar{G}_{ij}(z,t) = \bar{G}_{ij}(z, 2\delta - t)$. Hence, the identification of $\Sigma \times (2\delta)$ with $\Sigma \in$ the second copy of (M^3_Σ, g) is smooth by symmetry. This completes the smooth construction of the metric $(\bar{M}^3_{\Sigma,\delta}, \bar{g}_\delta)$.

Now define $H(z,t) = \Sigma_j h_j^j(z,t)$ to be the mean curvature of $\Sigma \times t$ in $(\Sigma \times [0,2\delta], G)$, and let $\dot{H}(z,t) = \frac{d}{dt}H(z,t)$. We note that

$$H(z,0) = 0 = H(z,2\delta) \tag{39}$$

since Σ is a horizon (and hence has zero mean curvature) in (M^3_Σ, g). Let $\alpha = \sup_z \dot{H}(z,0)$ and let $\beta = \sup_z \sum_{jk} h_j^k(z,0)h_k^j(z,0)$, which we note are functions of the metric g on M^3_Σ and are independent of δ. Then we require that the smooth extension we choose for h_j^k satisfies

$$\dot{H}(z,t) \le 2|\alpha| + 1, \tag{40}$$

(which is possible because of equation 39) and

$$\sum_{jk} h_j^k(z,t)h_k^j(z,t) \le 2\beta + 1. \tag{41}$$

Then combining equations 39 and 40 also yields

$$|H(z,t)| \le (2|\alpha| + 1)\delta. \tag{42}$$

These estimates allow us to bound the scalar curvature of $(\tilde{M}^3_{\Sigma,\delta}, \bar{g}_\delta)$ from below since by the second variation formula and the Gauss equation we have that

$$R = -2\dot{H} + 2K - |h|^2 - H^2 \tag{43}$$

where $R(z,t)$ is scalar curvature and $K(z,t)$ is the Gauss curvature of $\Sigma \times t$. At this point we realize that we also need a lower bound K_0 for $K(z,t)$ which is independent of δ, which follows from imposing an upper bound on the C^2 norm (in the z variable) of our smooth choice of $h_j^k(z,t)$. Then using this combined with inequalities 40, 41, and 42 we get

$$R(z,t) \ge R_0 \tag{44}$$

where R_0 is independent of δ (for $\delta < 1$).

Now we are ready to define $\tilde{g}_\delta$ using equation 37. We already know that $(\tilde{M}^3_{\Sigma,\delta}, \bar{g}_\delta)$ is smooth and has nonnegative scalar curvature everywhere except possibly in $\Sigma \times [0,2\delta]$ where it has $R \ge R_0$. If $R_0 \ge 0$, then we just let $u_\delta(x) = 1$ so that $\tilde{g} = \bar{g}$. Otherwise, we define $u_\delta(x)$ such that

$$(-8\Delta_{\bar{g}} + \mathcal{R}_\delta(x))u_\delta(x) = 0 \tag{45}$$

and $u_\delta(x)$ goes to one in both asymptotically flat ends, where $\mathcal{R}_\delta(x)$ equals R_0 in $\Sigma \times [0,2\delta]$, equals zero for x more than a distance δ from $\Sigma \times [0,2\delta]$, is smooth, and takes values in $[R_0, 0]$ everywhere. Then it follows that for sufficiently small δ, $u_\delta(x)$ is a smooth superharmonic function. Furthermore, since $\mathcal{R}_\delta$ is zero everywhere except on an open set whose volume is going to zero as δ goes to zero, and since $\mathcal{R}_\delta$ is uniformly bounded from below on this small set, it follows from bounding Green's functions from above that

$$1 \le u_\delta(x) \le 1 + \epsilon(\delta) \tag{46}$$

where ϵ goes to zero as δ approaches zero.

Furthermore, by equations 44, 37, and 95, $(\tilde{M}^3_{\Sigma,\delta}, \tilde{g}_\delta)$ has nonnegative scalar curvature, and since $u_\delta(x)$ and and $(\tilde{M}^3_{\Sigma,\delta}, \bar{g}_\delta)$ are smooth, $(\tilde{M}^3_{\Sigma,\delta}, \tilde{g}_\delta)$ is too. In addition, it follows from the construction of $(\tilde{M}^3_{\Sigma,\delta}, \tilde{g}_\delta)$ that there exists a diffeomorphism into $(\bar{M}^3, \bar{g})$ with respect to which the metrics are arbitrarily uniformly close to each other in the limit as δ goes to zero. Hence, by equation 46, we see that the same statement is true for $(\tilde{M}^3_{\Sigma,\delta}, \tilde{g}_\delta)$. Finally, it follows from equation 45 that $\tilde{m}_\delta$, the mass of $(\tilde{M}^3_{\Sigma,\delta}, \tilde{g}_\delta)$, converges to m, the mass of $(\bar{M}^3_\Sigma, \bar{g})$, in the limit as δ goes to zero. Hence, inequality 35 follows, proving inequality 32.

To prove the case of equality, we note that we can view the above proof in a different way. Since the singular manifold $(\bar{M}^3_\Sigma, \bar{g})$ is the limit of the smooth manifolds $(\tilde{M}^3_{\Sigma,\delta}, \tilde{g}_\delta)$ which have nonnegative scalar curvature, it follows that $(\bar{M}^3_\Sigma, \bar{g})$ has generalized nonnegative scalar curvature as defined in [**7**]. Then if we reexamine the proof of theorem 5, we see that the theorem, including the case of equality, is also true for singular manifolds like $(\bar{M}^3_\Sigma, \bar{g})$ which have generalized nonnegative scalar curvature (see the discussion at the end of the proof of theorem 5). Hence, by equation 34 and theorem 5, we get equality in inequality 32 if and only if $(\bar{M}^3_\Sigma, \bar{g})$ has zero scalar curvature and is conformal to $(\mathbf{R}^3, \delta)$ minus a finite number of points.

Since $(\bar{M}^3_\Sigma, \bar{g})$ has two ends, it must be conformal to $(\mathbf{R}^3 \backslash \{0\}, \delta)$, and since it has zero scalar curvature, it follows from equation 95 that it must be a Schwarzschild metric. Hence, in the case of equality for inequality 32, (M^3, g) must be a Schwarzschild manifold outside Σ. □

2.3. Proof That $m(t)$ Is Nonincreasing. In this subsection we will prove that $m(t)$, the total mass of (M^3, g_t), is non-increasing in t, as claimed in theorem 3. The fact that $m(t)$ is nonincreasing is of course central to the argument presented in this section for proving the Riemannian Penrose conjecture and is perhaps the most important property of the conformal flow of metrics $\{g_t\}$.

Recalling definition 9 from section 2.1, we begin by stating a lemma proved in [**6**].

LEMMA 2. *The left and right hand derivatives* $\frac{d}{dt^\pm}$ *of* $u_t(x)$ *exist for all* $t > 0$ *and are equal except at a countable number of* t*-values. Furthermore,*

$$\frac{d}{dt^+} u_t(x) = v_t^+(x) \tag{47}$$

and

$$\frac{d}{dt^-} u_t(x) = v_t^-(x) \tag{48}$$

where $v_t^\pm(x)$ *equals zero inside* $\Sigma^\pm(t)$ *(see definition 9) and outside* $\Sigma^\pm(t)$ *is the harmonic function which equals* 0 *on* $\Sigma^\pm(t)$ *and goes to* $-e^{-t}$ *at infinity.*

We will use this lemma to compute the left and right hand derivatives of $m(t)$. As proven at the end of appendix E, the flow of metrics $\{g_t\}$ we are considering has the property that the rate of change of the metric g_t is just a function of g_t and not of t or g_0. Hence, we will just prove that $m'(0) \leq 0$, from which it will follow that $m'(t) \leq 0$. So without loss of generality, we will assume that the flow begins at some time $-t_0 < 0$, and then compute the left and right hand derivatives of $m(t)$ at $t = 0$.

Also, in [**6**] we prove that $\Sigma^+(t)$ and $\Sigma^-(t)$ are horizons in (M^3, g_t). Furthermore, $v_0^\pm(x)$ is harmonic in (M^3, g_0), equals 0 on $\Sigma^\pm(0)$, and goes to -1 at infinity. Hence, by equation 31 and theorem 6 of the previous subsection,

$$v_0^\pm(x) = -1 + \frac{\mathcal{E}(\Sigma^\pm(0), g_0)}{2|x|} + \mathcal{O}\left(\frac{1}{|x|^2}\right) \tag{49}$$

where

$$m(0) \geq \frac{1}{2}\mathcal{E}(\Sigma^{\pm}(0), g_0). \tag{50}$$

Now we are ready to compute $m'(t)$. As in section 1.2, let $g_0 = \mathcal{U}_0(x)^4 g_{flat}$, where (M^3, g_{flat}) is isometric to $(\mathbf{R}^3 \backslash B_r(0), \delta)$ in the harmonically flat end, where we have chosen r and scaled the harmonically flat coordinate chart such that $\mathcal{U}_0(x)$ goes to one at infinity. Then by definition 2 for the total mass, we have that

$$\mathcal{U}_0(x) = 1 + \frac{m(0)}{2|x|} + \mathcal{O}\left(\frac{1}{|x|^2}\right). \tag{51}$$

We will also let $g_t = \mathcal{U}_t(x)^4 g_{flat}$ in the harmonically flat end. Then since $g_t = u_t(x)^4 g_0$, it follows that

$$\mathcal{U}_t(x) = u_t(x)\mathcal{U}_0(x). \tag{52}$$

Now we define $\alpha(t)$ and $\beta(t)$ such that

$$u_t(x) = \alpha(t) + \frac{\beta(t)}{|x|} + \mathcal{O}\left(\frac{1}{|x|^2}\right). \tag{53}$$

Then since $u_0(x) \equiv 1$ and $\frac{d}{dt^{\pm}} u_t(x)|_{t=0} = v_0^{\pm}(x)$, it follows from equation 49 that

$$\begin{aligned} \alpha(0) &= 1, & \tfrac{d}{dt^{\pm}}\alpha(t)|_{t=0} &= -1, \\ \beta(0) &= 0, & \tfrac{d}{dt^{\pm}}\beta(t)|_{t=0} &= \tfrac{1}{2}\mathcal{E}(\Sigma^{\pm}(0), g_0). \end{aligned} \tag{54}$$

Thus, by equation 52,

$$\mathcal{U}_t(x) = \alpha(t) + \frac{1}{|x|}\left(\beta(t) + \frac{m(0)}{2}\alpha(t)\right) + \mathcal{O}\left(\frac{1}{|x|^2}\right) \tag{55}$$

so that by definition 2

$$m(t) = 2\alpha(t)\left(\beta(t) + \frac{m(0)}{2}\alpha(t)\right). \tag{56}$$

Hence, by equation 54

$$\frac{d}{dt^{\pm}} m(t)|_{t=0} \;=\; \mathcal{E}(\Sigma^{\pm}(0), g_0) - 2m(0) \;\leq\; 0 \tag{57}$$

by equation 50. Then since we were able to choose $t = 0$ without loss of generality as previously discussed, we have proven the following theorem.

THEOREM 7. *The left and right hand derivatives $\frac{d}{dt^{\pm}}$ of $m(t)$ exist for all $t > 0$ and are equal except at a countable number of t-values. Furthermore,*

$$\frac{d}{dt^{\pm}} m(t) \leq 0 \tag{58}$$

for all $t > 0$ (and the right hand derivative of $m(t)$ at $t = 0$ exists and is nonpositive as well).

Hence, $m'(t)$ exists almost everywhere, and since $m(t)$ is continuous and its left and right hand derivatives are all nonpositive, $m(t)$ is nonincreasing.

2.4. Generalization to Asymptotically Flat Manifolds and the Case of Equality. Up to this point we have assumed that (M^3, g_0) was harmonically flat at infinity. In particular, theorems 2, 3, and 4 only apply to harmonically flat manifolds as stated. In this subsection, we will extend theorems 2 and 3 and elements of theorem 4 to asymptotically flat manifolds. This will prove the main theorem, theorem 1, except for the case of equality, which we will see follows from the case of equality of theorem 6.

It is worth noting that the main reason for initially considering only harmonically flat manifolds was convenience. Alternatively, we could have ignored harmonically flat manifolds and dealt only with asymptotically flat manifolds. However, this would have complicated some of the arguments unnecessariy, so we chose to delay these considerations until now.

DEFINITION 13. *(M^n, g) is said to be* **asymptotically flat** *if there is a compact set $K \subset M$ such that $M \backslash K$ is the disjoint union of ends $\{E_k\}$, such that for each end there exists a diffeomorphism $\Phi_k : E_k \to \mathbf{R}^n \backslash B_1(0)$ such that, in the coordinate chart defined by Φ_k,*

$$g = \sum_{i,j} g_{ij}(x) dx^i dx^j$$

where

$$g_{ij}(x) = \delta_{ij} + O(|x|^{-p})$$

$$|x||g_{ij,k}(x)| + |x|^2 |g_{ij,kl(x)}| = O(|x|^{-p})$$

$$|R(g)| = O(|x|^{-q})$$

for some $p > \frac{n-2}{2}$ and some $q > n$, where we have used commas to denote partial derivatives in the coordinate chart, and $R(g)$ is the scalar curvature of (M^n, g).

These assumptions on the asymptotic behavior of (M^n, g) at infinity imply the existence of the limit

$$M_{ADM}(g) = (4\omega_{n-1})^{-1} \lim_{\sigma \to \infty} \int_{S_\sigma} \sum_{i,j} (g_{ij,i}\nu_j - g_{ii,j}\nu_j)\, d\mu \tag{59}$$

where $\omega_{n-1} = Vol(S^{n-1}(1))$, S_σ is the coordinate sphere of radius σ, ν is the unit normal to S_σ, and $d\mu$ is the area element of S_σ in the coordinate chart. The quantity M_{ADM} is called the **total mass** of (M^n, g) (see [**1**], [**2**], [**29**], and [**33**]), and agrees with the definition of total mass for harmonically flat 3-manifolds given in definition 2.

It turns out that the arguments in the proof given in [**6**] of the existence theorem, theorem 2, do not use harmonic flatness anywhere, so we immediately get existence of the conformal flow of metrics for asymptotically flat manifolds. Similarly, the arguments used to prove that $A(t)$ is constant still hold. Next we reexaming the proof of theorem 7 which proved that $m(t)$ was nonincreasing. The only modification we need to make is to use the more general definition for the total mass of an asymptotically flat manifold given by equation 59. It is then straight forward to check that equation 57 and hence theorem 7 are still true. Hence,

THEOREM 8. *Theorems 2 and 3 are true for asymptotically flat manifolds as well as harmonically flat manifolds.*

We choose not to extend theorem 4 to asymptotically flat manifolds, but conjecture that it is still true. Instead, we observe that we must still have

$$m(t) \geq \sqrt{\frac{A(t)}{16\pi}} \tag{60}$$

for asymptotically flat manifolds. Otherwise, given an asymptotically flat counterexample, we could use lemma 1 to perturb the manifold slightly making it harmonically flat at infinity such that it still violated equation 60. Then applying the conformal flow of metrics to this harmonically flat manifold would violate theorems 3 and 4, which is a contradiction. Setting $t = 0$ in inequality 60 then proves the Riemannian Penrose inequality for asymptotically flat manifolds.

The case of equality of theorem 1 then follows from equation 57 and theorem 6. If we have equality in the Riemannian Penrose inequality, then applying the conformal flow of metrics to this initial metric must also give equality in inequality 60 for all $t \geq 0$. Hence, the right hand derivative of $m(t)$ at $t = 0$ equals zero, so by equation 57,

$$\mathcal{E}(\Sigma^+(0), g_0) = 2m(0). \tag{61}$$

By definition 9 and equation 13, $\Sigma^+(0)$ is the outermost minimal area enclosure of Σ_0 in (M^3, g). Furthermore, by the case of equality of theorem 6, (M^3, g) is a Schwarzschild manifold outside $\Sigma^+(0)$. Hence, $\Sigma^+(0)$ is the outermost horizon of (M^3, g), so (M^3, g) is isometric to a Schwarzschild manifold outside their respective outermost horizons. This proves the case of equality for theorem 1 and consequently the Riemannian Penrose inequality as well.

The reader might also have noticed that none of the arguments in this section have used anything about the original manifold inside the original horizon Σ_0. Hence, we can generalize theorem 1 to the following.

THEOREM 9. *Let (M^3, g) be a complete, smooth 3-manifold with boundary which has nonnegative scalar curvature and a single asymptotically flat end with total mass m. Then if the boundary is an outer-minimizing horizon (with one or more components) of total area A,*

$$m \geq \sqrt{\frac{A}{16\pi}} \tag{62}$$

with equality if and only if (M^3, g) is isometric to a Schwarzschild manifold outside their respective outermost horizons.

3. The Huisken-Ilmanen Proof of the Penrose Conjecture Using Inverse Mean Curvature Flows

We will now describe the approach taken by Huisken and Ilmanen[**19**] to establish the Penrose conjecture in the slightly weaker form.

THEOREM 10. *Let M be a complete, connected, asymptotically flat 3-manifold with nonnegative scalar curvature. Assume that the boundary is compact and consists of minimal surfaces, and that M contains no other compact minimal surfaces. Then $m \geq 0$, and*

$$16\pi m^2 \geq |N| \tag{63}$$

where $|N|$ is the area of any connected component of ∂M. Equality holds if and only if M is one-half of the spatial Schwarzschild manifold.

The main difference between this theorem and the full Riemannian Penrose conjecture is that here the lower bound on the mass is given in terms of the largest of the areas of the boundary components, while in the full conjecture it is given in terms of the total area of the boundary components. On the other hand, **[19]** does establish the conjecture in great generality, and it precedes the work of Bray **[6]** which proves the full conjecture. Furthermore the strategy employed in the Huisken/Ilmanen proof is totally different from that of Bray, and it is likely that their approach will shed light on other related questions concerning scalar curvature in three dimensions; indeed we describe an application to the positivity of a quasi-local mass defined by Bartnik**[3]**. Their approach involves the study of an evolution equation for compact 2-surfaces in M. This approach has a long history, and we now proceed to describe this history and the work of **[19]**.

3.1. Quasi-Local Mass. A central problem in general relativity is to understand how to quantify the amount of mass inside a given region. In appendix C, we define, in the spherically symmetric case, a function $m(V)$ for a 2-sphere containing a volume V which is increasing as a function of V beyond the outermost minimal sphere. Furthermore, for large V, $m(V)$ equals the total mass of the manifold. Hence, it seems reasonable to say that the spherically symmetric sphere $\Sigma(V)$ defined in appendix C contains a mass $m(V)$. The function $m(V)$ should be called a quasi-local mass function.

Naturally we would like to define a quasi-local mass function which would measure the amount of mass inside any surface Σ which is the boundary of a region in any 3-manifold M^3 (with $R \geq 0$). We refer the reader to **[10]**, **[3]** ,**[11]**, and **[13]** for more discussion on this topic.

A simple definition of quasi-local mass was proposed by Hawking and is called the Hawking mass. Referring to appendix C, we recall that in the spherically symmetric case,

$$m(V) = \left(\frac{A(V)}{16\pi}\right)^{\frac{1}{2}} \left(1 - \frac{1}{16\pi} A(V) A'(V)^2\right) \tag{64}$$

It so happens that $A'(V) = H(V)$, where $H(V)$ is the mean curvature of the spherically symmetric sphere $\Sigma(V)$. Hence, one way to generalize equation 64 is to define

$$m(\Sigma) = \left(\frac{A}{16\pi}\right)^{\frac{1}{2}} \left(1 - \frac{1}{16\pi}\int_\Sigma H^2\right) \tag{65}$$

where A is the area of Σ and H is the mean curvature of Σ in M^3. It turns out that this definition of quasi-local mass, first defined by Hawking, has some very important properties, and seems to be especially natural in case the surface Σ is of least area for the volume which it bounds. As we shall see, the Huisken/Ilmanen approach employs this definition for surfaces in their evolution, and shows that it has the correct monotonicity properties.

Another definition of quasi-local mass which has the potential to be meaningful for more general surfaces was given by Bartnik**[3]**. The Bartnik mass is defined for any compact 3-manifold Ω with boundary and with nonnegative scalar curvature which can be embedded as a domain in an asymptotically flat manifold M with $R \geq 0$ in such a way that $M \setminus \Omega$ contains no compact minimal surfaces. The Bartnik mass $m_B(\Omega)$ is then defined to be the infimum of the ADM masses of all possible such M which contain an isometric embedding of Ω as a subdomain. The fact that $m_B(\Omega) \geq 0$ follows from the positive mass theorem, but it is not clear the this quantity is positive since it is an infimum taken over a very large set of manifolds. The work of Huisken/Ilmanen implies the following result.

THEOREM 11. *For any Ω for which the Batnik mass is defined, it is true that $m_B(\Omega) > 0$ unless Ω is flat (hence locally isometric to a domain in $\mathbf{R}^3$).*

3.2. Inverse Mean Curvature Flow and Geroch Monotonicity. In [**11**], Geroch showed that if M^3 has nonnegative scalar curvature, then the Hawking mass is nondecreasing when the surface Σ is flowed outward at a speed equal to the inverse of its mean curvature. In view of this, Jang and Wald proposed using the Hawking mass function to prove the Penrose inequality [**26**]. They suggested that we should let $\Sigma(0)$ be an outermost horizon and then to flow out using an inverse mean curvature flow to create a family of surfaces $\Sigma(t)$ flowing out to infinity. Since the Hawking mass function $m(t) = m(\Sigma(t))$ is nondecreasing as a function of t, we have

$$\lim_{t\to\infty} m(t) \geq m(0) \tag{66}$$

Furthermore $m(0) = \sqrt{\frac{A}{16\pi}}$, where A is the area of the outermost horizon $\Sigma(0)$, since horizons have zero mean curvature. Hence, Jang and Wald proposed a proof of the Penrose inequality which is naturally a generalization of the proof which works in the spherically symmetric case.

The main difficulty in making this formal argument rigorous is to prove the existence and regularity for the inverse mean curvature flow. Naturally, if the mean curvature of the surface ever went to zero or became negative, the flow could not exist, at least in this form. Indeed, in [**19**] it is observed that if the initial surface Σ_0 in $\mathbf{R}^3$ is taken to be the boundary of a small tubular neighborhood of a standard circle, then Σ_0 will flow outward for a short time resulting in a fattening of the torus. At some time the mean curvature must go to zero, and the flow must become singular. The main difficulty in this approach is how to define the flow after times at which it becomes singular. In the next section we discuss how Huisken and Ilmanen solved this problem.

3.3. The Huisken-Ilmanen Proof. Recall that the smooth flow which we are studying is the inverse mean curvature flow given by

$$\frac{\partial x}{\partial t} = H^{-1}\nu(x), \quad x \in \Sigma_t \tag{67}$$

with initial condition Σ_0 specified to be a slight inward perturbation of the boundary surface with $H > 0$. (The condition that M contains no other compact minimal surfaces implies that the boundary surfaces are local minima for area, and guarantees that it is possible to find nearby surfaces in M with $H > 0$.) In (67) ν denotes the inward normal vector to M (opposite to the direction of the mean curvature vector). As we have seen, there is little hope that solutions of this flow will remain smooth as surfaces in M and continuous in t. To resolve this difficulty (and others) [**19**] rewrite the flow as an equation for the level sets of a function $u(x)$. First if we assume that the level sets of u are smooth surfaces and $|\nabla u| \neq 0$, the equation becomes

$$\operatorname{div}_M \left(\frac{\nabla u}{|\nabla u|} \right) = |\nabla u|. \tag{68}$$

Instead of being an evolution equation, (68) is a degenerate elliptic equation. It is degenerate because the leading order part is invariant under the replacement of u by $f(u)$ since it measures the mean curvature of the level surfaces of u. The advantage of looking at a level set formulation of the flow is that it becomes possible for the level sets to "fatten", and in this way to jump in space. This is an essential feature of any weak formulation of the flow, as examples such as the torus example suggest.

Since we cannot expect solutions to be smooth, we may formulate a notion of weak solution. Since the left hand side of (68) is the Euler-Lagrange operator obtained from the functional $\int_M |\nabla u|\,dx$, it is natural to consider freezing the right hand side as $|\nabla u|$, and considering the functional $J_u(v) := \int_K (|\nabla v| + v|\nabla u|)\,dx$ where K is a compact subset of M. We may then seek minima of this functional for a given u and hope that for a suitable choice of u, the function u will achieve the minimum. In any event, we make the following definition.

DEFINITION 14. *A locally Lipschitz function u on an open set Ω is called a weak solution of (68) on Ω if*

$$J_u^K(u) \le J_u^K(v) \tag{69}$$

for every locally Lipschitz function v such that $u - v$ has compact support in Ω and K is any compact subset of Ω containing the support of $u - v$.

We can think of the level sets $E_t = \{x : u(x) < t\}$ as evolving, and one can impose initial conditions on the solutions by choosing an open set E_0. We say that u is a solution of the initial value problem with initial condition E_0 if u is a locally Lipschitz solution of (69) in $M \setminus E_0$, and $E_0 = \{u < 0\}$.

In order to come to a better geometric understanding of weak solutions, it is useful to reformulate the condition strictly in terms of level sets. This can be done in the following way. Given a locally Lipschitz function u, we may define the functional $J_u(\cdot)$ on measurable sets F by

$$J_u^K(F) = |\partial^* F \cap K| - \int_{F\cap K} |\nabla u| \tag{70}$$

where K is a compact subset of M and $|\partial^* F|$ denotes the 2-dimensional Hausdorff measure of the reduced boundary of F (assuming that F is a set of finite perimeter). We then say that a set E minimizes J_u in a set A if

$$J_u^K(E) \le J_u^K(F)$$

for any set F with symmetric difference $F \Delta E \subseteq K$, and K compact in A. The following result holds ([**19**]).

LEMMA 3. *A locally Lipschitz function u satisfies (69) in an open set Ω if and only if for each t, E_t minimizes J_u in Ω.*

The above lemma allows us to use known regularity results to obtain the following regularity results for solutions of (69) with intitial condition E_0. Precisely, if ∂E_0 is $C^{1,\alpha}$ for some $\alpha > 0$, then for each $t \ge 0$ we have ∂E_t is also $C^{1,\alpha}$; in fact, we also have $\partial\{u > t\}$ is $C^{1,\alpha}$. This will be important for establishing monotonicity of the Hawking mass.

Another important geometric property of weak solutions which is essential for proving the Penrose conjecture is that the sets E_t are *area minimizing hulls.* This should be thought of as a mean curvature analogue of convexity. If E is a set and Ω is an open set, we say that E is a *minimizing hull* in Ω if E minimizes boundary area for sets containing E. Precisely, if

$$|\partial^* E \cap K| \leq |\partial^* F \cap K|$$

for any F containing E such that $F \setminus E \subset K$ and K is a compact set of Ω.

The existence and uniqueness theorem for the initial value problem can be stated.

THEOREM 12. *Assume that M is an asymptotically flat 3-manifold without boundary. For any nonempty bounded smooth open set E_0, there exists a proper, locally Lipschitz solution of (69) satisfying the initial condition E_0. Furthermore this solution is unique in $M \setminus E_0$.*

This solution is constructed by an elliptic regularization procedure. The idea is to change the equation (68) to the following

$$\operatorname{div}_M \left(\frac{\nabla u^\epsilon}{\sqrt{|\nabla u^\epsilon|^2 + \epsilon^2}} \right) - \sqrt{|\nabla u^\epsilon|^2 + \epsilon^2} = 0. \tag{71}$$

This equation has the following geometric meaning. It is simply the equation which says that the family of hypersurfaces $N_t = \text{graph}(\epsilon^{-1}\ (u^\epsilon - t))$ is a solution of the inverse mean curvature flow in $M \times \mathbf{R}$. The approach of **[19]** is then to solve (71) with $\epsilon > 0$ small on $B_R \setminus E_0$ where R is chosen large, and the boundary conditions $u^\epsilon = 0$ on ∂E_0 and $u^\epsilon = R$ on ∂B_R are imposed. A uniform gradient estimate on such solutions is then derived and it is shown that the u^ϵ converge to a proper solution of (69) with initial conditions E_0.

Another important property of the flow is that if ∂E_0 is connected, then it is true that ∂E_t is connected for all $t > 0$ as well. This follows from the maximum principle as a disconnected level set could only occur is u had an interior maximum point. Furthermore, the large time asymptotics of the flow can be analyzed, and for large t it is shown that ∂E_t is smooth and sufficiently close to a large coordinate sphere that the Hawking mass converges to the ADM mass.

Finally, to prove the monotonicity of the Hawking mass under this weak flow, an analagous formula is derived for the smooth graphical flow given by the graph of u^ϵ, and a delicate limiting argument is used to show the monotonicity for the actual limiting flow. The formal argument used by Geroch, Jang, and Wald can then be applied for the Huisken/Ilmanen flow, and this establishes the Penrose conjecture in case there is a single horizon (boundary component of M).

To handle the case of several horizons, Huisken and Ilmanen run the flow starting from one chosen horizon, leaving the others fixed. They then show that there is no loss of Hawking mass at the times at which this flow jumps across the other horizons. This is done by flowing until E_t almost touches another horizon and then taking the minimizing hull of E_t and that horizon. They show that the Hawking mass does not decrease in this process, and the minimizing hull has connected boundary. In this way they engulf all of the horizons, and show that the Penrose inequality holds where the area is the maximum area of the horizons.

4. New Quasi-Local Mass Functions

In this section we define several new quasi-local mass functions following [6]. The idea of a quasi-local mass function is that it should define how much mass is in a given region (or, equivalently, inside a given surface) of (M^3, g). When the region is very large, the quasi-local mass function should approach the total mass of (M^3, g). In addition, a reasonable quasi-local mass function should have some kind of monotonicity property, meaning that if one region contains another region, then it should have larger mass.

In [19], Huisken and Ilmanen used the Hawking mass to prove the Riemannian Penrose inequality for a single horizon. In fact, they proved that the Hawking mass of a surface is nondecreasing with respect to their generalized inverse mean curvature flow. With this in mind, it is not too surprising that the other proof the Riemannian Penrose inequality given in section 2 motivates yet another definition of quasi-local mass $m_g(\Sigma)$ (definition 15). This quasi-local mass function is also nondecreasing with respect to a certain family of surfaces, which we will describe. We then end the section by defining two more new quasi-local mass functions, $\tilde{m}_g(\Sigma)$ and $m_{inner}(\Sigma)$, which have very good monotonicity properties but are typically impractical to compute.

Let (M^3, g) be a complete asymptotically flat manifold with nonnegative scalar curvature and total mass m. Let Σ be any surface in M^3 which is in the class of surfaces $\mathcal{S}$ defined in section 1.2.

DEFINITION 15. *Suppose $u(x)$ is a positive harmonic function in (M^3, g) outside Σ going to a constant at infinity scaled such that $(M^3, u(x)^4 g)$ has the same total mass as (M^3, g).*

Then if Σ is an outer-minimizing horizon with area A in $(M^3, u(x)^4 g)$, we define the quasi-local mass of Σ in (M^3, g) to be

$$m_g(\Sigma) = \sqrt{\frac{A}{16\pi}}. \tag{72}$$

DEFINITION 16. *We define S to be the subset of $\mathcal{S}$ of surfaces Σ for which such a conformal factor $u(x)$ exists, and we note that (by equation 94 mostly) this conformal factor is unique for each Σ when it exists.*

As usual Σ could have multiple components. It is also interesting that

$$m_g(\tilde{\Sigma}) = m_{u(x)^4 g}(\tilde{\Sigma}) \tag{73}$$

for all surfaces $\tilde{\Sigma} \in S$ where the conformal factor $u(x)$ is any harmonic function in (M^3, g) defined outside $\tilde{\Sigma}$ which goes to a constant at infinity scaled such that $(M^3, u(x)^4 g)$ has the same total mass as (M^3, g).

LEMMA 4. *The quasi-local mass function $m_g(\Sigma)$ defined for (M^3, g) is monotone increasing for the family of surfaces $\Sigma(t)$ defined by equation 13. That is, $m_g(\Sigma(t))$ is nondecreasing in t. Furthermore,*

$$m_g(\Sigma(0)) = \sqrt{\frac{A}{16\pi}} \tag{74}$$

where A is the area of the original outer-minimizing horizon Σ_0 in (M^3, g), and

$$\lim_{t\to\infty} m_g(\Sigma(t)) = m, \tag{75}$$

the total mass of (M^3, g).

Proof. We consider the conformal flow of metrics g_t beginning with (M^3, g) defined in section 2. Then we note that by equation 13, $\Sigma(t)$ is an outer-minimizing horizon in $(M^3, u_t(x)^4 g)$ with area $A(t)$. Hence, $u_t(x)$ satisfies the conditions in definition 15 except that it is not scaled to have the correct mass. Hence, since mass has units of length, it follows that

$$m_g(\Sigma(t)) = \frac{m}{m(t)}\sqrt{\frac{A(t)}{16\pi}} \tag{76}$$

where again m is the total mass of (M^3, g). Then the lemma follows from theorems 3 and 4 and the fact that $m(0) = m$.

There is a trick which allows us to extend the definition of this quasi-local mass function to all surfaces in $\mathcal{S}$.

DEFINITION 17. *Define*

$$\tilde{m}_g(\Sigma) = \sup\left\{m_g(\tilde{\Sigma}) \mid \Sigma \text{ (entirely) encloses } \tilde{\Sigma} \in \mathcal{S}\right\} \tag{77}$$

where Σ is any surface in $\mathcal{S}$.

It follows trivially that $\tilde{m}_g(\Sigma)$ is monotone with respect to enclosure, which is a desirable property for quasi-local mass functions to have since larger regions should contain more mass in the nonnegative energy density setting which we are in. We also note that this same construction can be used with the Hawking mass to make it monotone with respect to enclosure too, where the original Hawking mass should only be defined to exist for surfaces which equal their own outermost minimal area enclosures, as motivated by the results of Huisken and Ilmanen in [**19**].

We also notice that the existence of the Penrose inequality allows us to define another new quasi-local mass function which is similar in nature to the Bartnik mass [**3**]. In fact, whereas the Bartnik mass could also be called an outer quasi-local mass function, it makes sense to call the new quasi-local mass function defined below the inner quasi-local mass function, which is clear from the definition.

DEFINITION 18. *Given a surface $\Sigma \in \mathcal{S}$ in (M^3, g), consider all other asymptotically flat, complete, Riemannian manifolds $(\tilde{M}^3, \tilde{g})$ with nonnegative scalar curvature which are isometric to (M^3, g) outside Σ. Then we define*

$$m_{inner}(\Sigma) = \sup\sqrt{\frac{\tilde{A}}{16\pi}} \tag{78}$$

where $\tilde{A}$ is the infimum of the areas of all of the surfaces in $(\tilde{M}^3, \tilde{g})$ in $\tilde{\mathcal{S}}$.

We note here that $\tilde{\mathcal{S}}$ is defined the same way as $\mathcal{S}$ in definition 3 and that the surface in $\tilde{\mathcal{S}}$ with minimum area may have multiple components. We also note that for $\tilde{A}$ to be nonzero that $(\tilde{M}^3, \tilde{g})$ must have more than one asymptotically flat end.

LEMMA 5. *Let (M^3, g) be an asymptotically flat, complete, Riemannian manifold with nonnegative scalar curvature, and let $\Sigma_1, \Sigma_2 \in \mathcal{S}$ such that Σ_2 (entirely) encloses Σ_1. Then*

$$m \geq m_{inner}(\Sigma_2) \geq m_{inner}(\Sigma_1) \tag{79}$$

where m is the total mass of (M^3, g).

Proof. Follows directly from the Penrose inequality and definition 18.

Also, if Σ is outer-minimizing (see definition 6), then

$$m_{outer}(\Sigma) \geq m_{inner}(\Sigma) \tag{80}$$

where $m_{outer}(\Sigma)$ is basically the Bartnik mass [**3**] except that we only consider extensions of the metric in which Σ continues to be outer-minimizing. The proof of this inequality and related discussions will be included in a paper on quasi-local mass which is currently in progress.

Appendix A. Motivation behind the Penrose Conjecture

In 1973, Roger Penrose proposed the Penrose inequality as a test of the cosmic censor hypothesis [**27**]. The cosmic censor hypothesis states that naked singularities do not develop starting with physically reasonable nonsingular generic initial conditions for the Cauchy problem in general relativity. (However, it has been shown by Christodoulou [**9**] that naked singularities can develop from nongeneric initial conditions.) If naked singularities did typically develop from generic initial conditions, then this could be a serious problem for general relativity as a theory since it might not be possible to solve the Einstein equations uniquely past these singularities. Singularities such as black holes do develop but are shielded from observers at infinity by their horizons so that the Einstein equations can still be solved from the point of view of an observer at infinity.

A summary of Penrose's argument can be found in [**26**]. The main idea is to consider a space-time $(N^4, \bar{g})$ with given initial conditions for the Cauchy problem (M^3, g, h) which satisfy the constraint equations given in equations 2, 3, and 4. To get the Riemannian Penrose conjecture, we make the additional assumption that the second fundamental form $h \equiv 0$, which by the constraint equations implies that (M^3, g) has nonnegative scalar curvature. We also suppose that (M^3, g) has outermost apparent horizons of total area A, event horizons of total area A_i, and total mass m_i (see [**16**], [**17**] for the definitions of these horizons). As long as a singularity does not form, then it is assumed that eventually the space-time should converge on some stationary final state, presumably a Kerr solution ([**21**], [**15**], [**28**]), which satisfies

$$A_f = 8\pi[m_f^2 + (m_f^4 - J^2)^{\frac{1}{2}}] \leq 16\pi m_f^2, \tag{81}$$

where A_f is the area of the horizon of the Kerr black hole, m_f is the mass at infinity, and J is the angular momentum.

However, by the Hawking area theorem [**14**], the total area of the event horizons of the black holes is nondecreasing. Thus, $A_f \geq A_i$. Also, presumably some energy radiates off to infinity, so we expect to have $m_i \geq m_f$.

Apparent horizons are defined to be the outer boundaries of regions in M^3 which contain trapped or marginally trapped surfaces [**16**]. Apparent horizons themselves must then be marginally trapped surfaces, and hence satisfy

$$H + h^{ij}(g_{ij} - r_i r_j) = 0 \tag{82}$$

where H is the mean curvature of the apparent horizons in M^3, h is the second fundamental form of (M^3, g) in $(N^4, \bar{g})$, and r is the outward unit normal to the apparent horizons in M^3. Hence, since we chose M^3 to have zero second fundamental form, the apparent horizons are zero mean curvature surfaces in M^3. Furthermore, if we consider the surface (possibly with multiple components) of smallest area which encloses the apparent horizons, it too must have zero mean curvature and hence is a marginally trapped surface in M^3. Thus, the union of the apparent horizons in M^3 form an outermost minimal surface of M^3. Furthermore, we note that it follows from a stability argument [**16**] that each component of this outermost minimal surface must be a sphere. Since event horizons always contain the apparent horizons, $A_i \geq A$, so putting all the inequalities together we conclude that

$$m_i \geq m_f \geq \sqrt{\frac{A_f}{16\pi}} \geq \sqrt{\frac{A_i}{16\pi}} \geq \sqrt{\frac{A}{16\pi}} \tag{83}$$

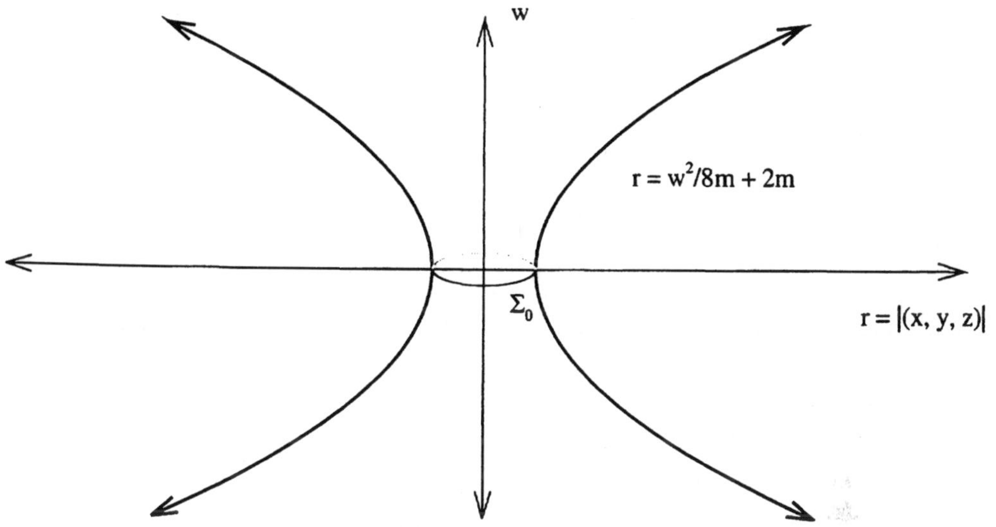

FIGURE 1. The space-like Schwarzschild metric of mass m, $(\mathbf{R}^3 - \{0\}, h)$, viewed as a submanifold of four-dimensional Euclidean space.

Thus, Penrose argued, assuming the cosmic censor hypothesis and a few reasonable sounding assumptions about the nature of gravitational collapse, given a complete asymptotically flat 3-manifold M^3 of total mass m_i with nonnegative scalar curvature which has an outermost minimal surface of total area A, then

$$m_i \geq \sqrt{\frac{A}{16\pi}} \tag{84}$$

Conversely, he argued, if one could find an M^3 which was a counterexample to the above inequality, then it would be likely that the counterexample, when used as initial conditions in the Cauchy problem for Einstein's equation, would produce a naked singularity. Hence, the two proofs of the above inequality given in sections 2 and 3 (for a single horizon) rule out one possibility for forcing counterexamples to the cosmic censor hypothesis.

Appendix B. The Schwarzschild Metric

The space-like Schwarzschild metric is a particularly important example to consider when discussing the Penrose inequality. First of all, it is the only 3-manifold which gives equality in the Penrose inequality. Also, if a 3-manifold is assumed to be complete, spherically symmetric, and have zero scalar curvature, then it must be isometric to either a Schwarzschild metric of mass $m > 0$ or $\mathbf{R}^3$, which can be viewed as the Schwarzschild metric when $m = 0$.

The space-like Schwarzschild metric, $(\mathbf{R}^3 - \{0\}, h)$, is a time symmetric, asymptotically flat, three-dimensional maximal slice (chosen to have zero momentum at infinity) of the four-dimensional Schwarzschild space-time metric.The space-like Schwarzschild metric is conformal to $\mathbf{R}^3 - \{0\}$ with $h_{ij} = \left(1 + \frac{m}{2r}\right)^4 \delta_{ij}$. The Schwarzschild metric of mass m, $(\mathbf{R}^3 - \{0\}, h)$, can also be isometrically embedded into four-dimensional Euclidean space as the three-dimensional set of points in $\mathbf{R}^4 =$

$\{(x, y, z, w)\}$ satisfying $|(x, y, z)| = \frac{w^2}{8m} + 2m$, seen in figure 1. Hence, Σ_0 is a minimal sphere of area $16\pi m^2$, so we have equality in the Penrose inequality.

Appendix C. The Spherically Symmetric Case

In this section we sketch a proof of the Penrose inequality in the case that M^3 is spherically symmetric. The proof is very easy conceptually, but what is more important is that some of the ideas generalize. In particular, we will see why the minimal sphere in the Penrose inequality must be outermost.

Let (M^3, g) be a complete, spherically symmetric 3-manifold with nonnegative scalar curvature which has a single, asymptotically flat end. For convenience, we also assume that (M^3, g) is isometric to the Schwarzschild metric of some mass m outside a large compact set. Then the total mass of (M^3, g) is m. Let $\Sigma(V)$ be the spherically symmetric sphere containing a volume V in M^3. Let $A(V)$ be the area of this sphere. It turns out that the function $A(V)$, $V \geq 0$, captures all the information about M^3 since M^3 is spherically symmetric.

Let $R(V)$ be the scalar curvature of M^3 on $\Sigma(V)$. It happens that

$$R(V) = \frac{8\pi}{A} - 2A(V)A''(V) - \frac{3}{2}A'(V)^2. \tag{85}$$

Define

$$m(V) = \left(\frac{A(V)}{16\pi}\right)^{\frac{1}{2}} \left(1 - \frac{1}{16\pi}A(V)A'(V)^2\right). \tag{86}$$

It turns out that $m'(V) \geq 0$ whenever $A'(V) \geq 0$ since

$$m'(V) = \frac{A'(V)}{16\pi}\left(\frac{A(V)}{16\pi}\right)^{\frac{1}{2}} R(V) \tag{87}$$

and $R(V) \geq 0$.

Let $\Sigma(V_0)$ be the outermost minimal sphere. It follows that $A'(V) \geq 0$ for all $V \geq V_0$. Hence, $m(V)$ is increasing in this range as well, so

$$\lim_{V \to \infty} m(V) \geq m(V_0) \tag{88}$$

Furthermore $m(V_0) = \sqrt{\frac{A(V_0)}{16\pi}}$ since $A'(V_0) = 0$. Also, we assumed that M^3 was isometric to the Schwarzschild metric outside a large compact set, and we claim that $m(V) = m$, the mass parameter of the Schwarzschild metric, in this region, or equivalently, for $V > V_{LARGE}$ for some $V_{LARGE} > 0$. To see this, consider the mass function $m(V)$ defined on the Schwarzschild metric, where now V refers to the volume contained by the spherically symmetric spheres of the Schwarzschild metric which is outside the horizon. Then by equation 87, $m(V)$ is constant for all V on the Schwarzschild metric since the Schwarzschild metric has zero scalar curvature. Furthermore, setting $V = 0$ and considering $m(V)$ at the horizon yields

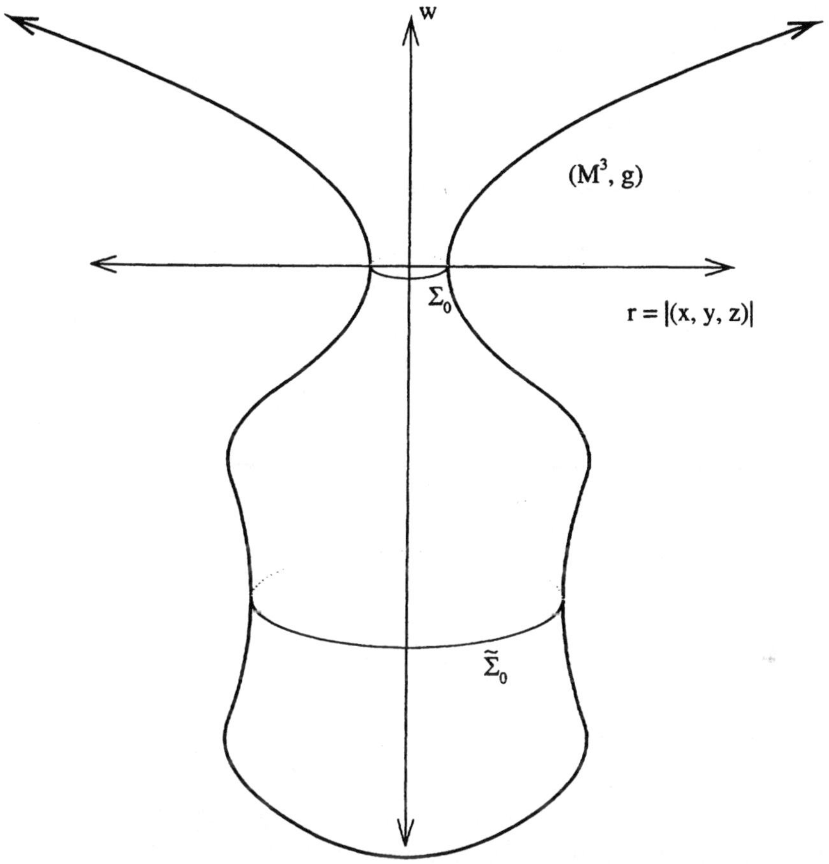

FIGURE 2. Counterexample to Penrose inequality if the minimal sphere is not outermost.

$m(0) = \sqrt{\frac{A(0)}{16\pi}} = m$, the mass parameter of the Schwarzschild metric, since the Schwarzschild metric gives equality in the Penrose inequality. Thus, $m(V) = m$ for all V in the Schwarzschild metric, so going back to (M^3, g), we see that $m(V) = m$, the mass parameter of the Schwarzschild metric, for $V > V_{LARGE}$. Thus, it follows from inequality 88 that

$$m \geq \sqrt{\frac{A(V_0)}{16\pi}} \tag{89}$$

which proves the Penrose inequality for spherically symmetric manifolds.

Conversely, we see that equation 87 can be used to construct spherically symmetric manifolds which do not satisfy the Penrose inequality if we do not require the minimal sphere to be outermost. In figure 2, we are viewing (M^3, g) as an isometrically embedded submanifold of $\mathbf{R}^4$ with the standard Euclidean metric. (M^3, g) is spherically symmetric and is constructed by rotating the curve shown above around the w-axis in $\mathbf{R}^4$. Hence, Σ_0 and $\tilde{\Sigma}_0$ are both 2-spheres, and we can choose the curve shown above so that the scalar curvature of (M^3, g), $R(g)$, is non-negative.

The Penrose inequality, $m \geq \sqrt{\frac{|\Sigma_0|}{16\pi}}$, is true for Σ_0, but is not true for $\tilde{\Sigma}_0$. However, $\tilde{\Sigma}_0$ is not an outermost minimal sphere since it is contained by another minimal sphere, namely, Σ_0. In fact, since $\tilde{\Sigma}_0$ is not outermost, we can construct a spherically symmetric manifold like the one shown above so that the area of $\tilde{\Sigma}_0$ is as large as we like and the total mass of (M^3, g) is still one.

Appendix D. An Example Solution to the Conformal Flow of Metrics

In this appendix we give the simplest example of a solution to the first order o.d.e. conformal flow of metrics defined by equations 12, 13, 14, and 15. The initial metric in this example is the three dimensional, space-like Schwarzschild metric which represents a single, non-rotating black hole in vacuum. The Schwarzschild metrics are also very natural from a geometric standpoint as well since they are spherically symmetric and have zero scalar curvature.

Since the flow does not change the metric inside the horizon, we will define this metric to have its horizon as a boundary, which is always allowable. Then (M^3, g_0) will be defined to be isometric to $(\mathbf{R}^3 - B_{m/2}(0), \mathcal{U}_0(x)^4 \delta_{ij})$, where

$$\mathcal{U}_0(x) = 1 + \frac{m}{2r} \tag{90}$$

where r is the distance from the origin in $(\mathbf{R}^3, \delta_{ij})$ and m is a positive constant equal to the mass of the black hole.

Next we define (M^3, g_t) to be isometric to $(\mathbf{R}^3 - B_{m/2}(0), \mathcal{U}_t(x)^4 \delta_{ij})$, where

$$\mathcal{U}_t(x) = \begin{cases} e^{-t} + \frac{m}{2r} e^t, & \text{for } r \geq \frac{m}{2} e^{2t} \\ \sqrt{\frac{2m}{r}}, & \text{for } r < \frac{m}{2} e^{2t}. \end{cases} \tag{91}$$

We note that on this metric the outermost horizon (and also the outermost minimal area enclosure of the original horizon) is the coordinate sphere given by $r = \frac{m}{2} e^{2t}$, so by equation 13 we define this horizon to be $\Sigma(t)$.

Next we recall from equation 12 that $g_t = u_t(x)^4 g_0$. Hence, $u_t(x) = \mathcal{U}_t(x)/\mathcal{U}_0(x)$. Furthermore, by equation 15 we must have $v_t(x) = \frac{d}{dt} u_t(x)$, so

$$v_t(x) = \frac{1}{\mathcal{U}_0(x)} \begin{cases} -e^{-t} + \frac{m}{2r} e^t, & \text{for } r \geq \frac{m}{2} e^{2t} \\ 0, & \text{for } r < \frac{m}{2} e^{2t}. \end{cases} \tag{92}$$

By equation 94, $v_t(x)$ is harmonic on (M^3, g_0) outside $\Sigma(t)$ since $a + b/r$ is harmonic in $(\mathbf{R}^3, \delta_{ij})$. Then since $v_t(x)$ goes to $-e^{-t}$ at infinity, is continuous, and equals zero inside $\Sigma(t)$, it follows that equation 14 is satisfied. Hence, (M^3, g_t) is a solution to the first order conformal flow of metrics defined by equations 12, 13, 14, and 15.

This example is a good example to keep in mind when considering the main theorems of section 2. For example, we notice that by definition 2 the total mass $m(t)$ of (M^3, g_t) equals m and hence is nonincreasing as claimed in section 2.3, and the area $A(t)$ of the horizon $\Sigma(t)$ in (M^3, g_t) is constant as claimed in theorem 3. Also, we see that the diameter of $\Sigma(t)$ is growing exponentially and contains any given bounded set in a finite amount of time, which turns out to be a general characteristic of the flow and is proven in [**6**].

Finally, we note that for all $t \geq 0$ in this example, (M^3, g_t) is isometric to a Schwarzschild metric of total mass m outside their respective horizons. Hence, even though the metric is shrinking pointwise, it is not changing at all outside its horizon, *after a reparametrization of the metric.* It is in this sense that theorem 4 states that no matter what the initial metric is, it eventually converges to a Schwarzschild metric outside its horizon.

Appendix E. The Harmonic Conformal Class of a Metric

In this appendix we define a new equivalence class and partial ordering of conformal metrics following [**6**]. As we will see, the existence of this equivalence class of metrics provides a natural motivation for the techniques used in section 2 to prove the Riemannian Penrose conjecture.

Let

$$g_2 = u(x)^{\frac{4}{n-2}} g_1 \tag{93}$$

where g_2 and g_1 are metrics on an n-dimensional manifold M^n, $n \geq 3$. Then we get the surprisingly simple identity that

$$\Delta_{g_1}(u\phi) = u^{\frac{n+2}{n-2}} \Delta_{g_2}(\phi) + \phi \Delta_{g_1}(u) \tag{94}$$

for any smooth function ϕ.

This motivates us to define the following relation.

DEFINITION 19. *Define*

$$g_2 \sim g_1$$

if and only if equation 93 is satisfied with $\Delta_{g_1}(u) = 0$ *and* $u(x) > 0$.

Then from equation 94 we get the following lemma.

LEMMA 6. *The relation* $\sim$ *is reflexive, symmetric, and transitive, and hence is an equivalence relation.*

Thus, we can define the following equivalence class of metrics.

DEFINITION 20. *Define*

$$[g]_H = \{\bar{g} \mid \bar{g} \sim g\}$$

to be the **harmonic conformal class** *of the metric* g.

Of course, this definition is most interesting when (M^n, g) has nonconstant positive harmonic functions, which happens for example when (M^n, g) has a boundary.

Also, we can modify the relation $\sim$ to get another relation $\succeq$.

DEFINITION 21. *Define*

$$g_2 \succeq g_1$$

if and only if equation 93 is satisfied with $-\Delta_{g_1}(u) \geq 0$ *and* $u(x) > 0$.

Then from equation 94 we get the following lemma.

LEMMA 7. *The relation $\succeq$ is reflexive and transitive, and hence is a partial ordering.*

Since $\succeq$ is defined in terms of superharmonic functions, we will call it the superharmonic partial ordering of metrics on M^n. Then it is natural to define the following set of metrics.

DEFINITION 22. *Define*

$$[g]_S = \{\bar{g} \mid \bar{g} \succeq g\}.$$

This set of metrics has the property that if $\bar{g} \in [g]_S$, then $[\bar{g}]_S \subset [g]_S$

Also, the scalar curvature transforms nicely under a conformal change of the metric. In fact, assuming equation 93 again,

$$R(g_2) = u(x)^{-\left(\frac{n+2}{n-2}\right)} \left(-c_n \Delta_{g_1} + R(g_1)\right) u(x) \tag{95}$$

where $c_n = \frac{4(n-1)}{n-2}$. This gives us the following lemma.

LEMMA 8. *The sign of the scalar curvature is preserved pointwise by $\sim$. That is, if $g_2 \sim g_1$, then $sgn(R(g_2)(x)) = sgn(R(g_1)(x))$ for all $x \in M^n$.*

Also, if $g_2 \succeq g_1$, and g_1 has non-negative scalar curvature, then g_2 has non-negative scalar curvature.

Hence, the harmonic conformal equivalence relation $\sim$ and the superharmonic partial ordering $\succeq$ are useful for studying questions about scalar curvature. In particular, these notions are useful for studying the Riemannian Penrose inequality which concerns asymptotically flat 3-manifolds (M^3, g) with non-negative scalar curvature. Given such a manifold, define $m(g)$ to be the total mass of (M^3, g) and $A(g)$ to be the area of the outermost horizon (which could have multiple components) of (M^3, g). Define $P(g) = \frac{m(g)}{\sqrt{A(g)}}$ to be the Penrose quotient of (M^3, g). Then an interesting question is to ask which metric in $[g]_S$ minimizes $P(g)$.

Section 2 can be viewed as an answer to the above question. We defined a conformal flow of metrics (starting with g_0) for which the Penrose quotient was non-increasing, and in fact this conformal flow stays inside $[g_0]_S$. Furthermore, $g_{t_2} \in [g_{t_1}]_S$ for all $t_2 \geq t_1 \geq 0$. Also, no matter which metric we start with, the metric converges to a Schwarzschild metric outside its horizon **[6]**. Hence, the minimum value of $P(g)$ in $[g]_S$ is achieved in the limit by metrics converging to a Schwarzschild metric (outside their respective horizons).

In the case that the g is harmonically flat at infinity, a Schwarzschild metric (outside the horizon) is contained in $[g]_S$. More generally, given any asymptotically flat manifold (M^3, g), we can use $\mathbf{R}^3 \backslash B_r(0)$ as a coordinate chart for the asymptotically flat end of (M^3, g) which we are interested in, where the metric g_{ij} approaches δ_{ij} at infinity in this coordinate chart. Then we can consider the conformal metric

$$g_C = \left(1 + \frac{C}{|x|}\right)^4 g \tag{96}$$

in this end. In the limit as C goes to infinity, the horizon will approach the coordinate sphere of radius C. Then outside this horizon in the limit as C goes to infinity, the function $(1 + \frac{C}{|x|})$ will be close to a superharmonic function on (M^3, g) and the metric g_C will approach a Schwarzschild metric (since the metric g is approaching the standard metric on $\mathbf{R}^3$). Hence, the Penrose quotient of g_C will approach $(16\pi)^{-1/2}$, which is the Penrose quotient of a Schwarzschild metric.

As a final note, we prove that the first order o.d.e. for $\{g_t\}$ defined in equations 12,13, 14, and 15 is naturally defined in the sense that the rate of change of g_t is a function only of g_t and not of g_0 or t. To see this, given any solution $g_t = u_t(x)^4 g_0$ to equations 12,13, 14, and 15, choose any $s > 0$ and define $\bar{u}_t(x) = u_t(x)/u_s(x)$ so that

$$g_t = \bar{u}_t(x)^4 g_s \tag{97}$$

and $\bar{u}_s(x) \equiv 1$. Then define $\bar{v}_t(x)$ such that

$$\begin{cases} \Delta_{g_s} \bar{v}_t(x) \equiv 0 & \text{outside } \Sigma(t) \\ \bar{v}_t(x) = 0 & \text{on } \Sigma(t) \\ \lim_{x\to\infty} \bar{v}_t(x) = -e^{-(t-s)} & \end{cases} \tag{98}$$

and $\bar{v}_t(x) \equiv 0$ inside $\Sigma(t)$. Then what we want to show is

$$\bar{u}_t(x) = 1 + \int_s^t \bar{v}_r(x)dr \tag{99}$$

To prove the above equation, we observe that from equations 94, 98, and 14 it follows that

$$v_t(x) = \bar{v}_t(x)\, u_s(x) \tag{100}$$

since $\lim_{x\to\infty} u_s(x) = e^{-s}$. Hence, since

$$u_t(x) = u_s(x) + \int_s^t v_r(x)dr \tag{101}$$

by equation 15, dividing through by $u_s(x)$ yields equation 99 as desired. Thus, we see that the rate of change of $g_t(x)$ at $t = s$ is a function of $\bar{v}_s(x)$ which in turn is just a function of $g_s(x)$ and the horizon $\Sigma(s)$. Hence, to understand properties of the flow we need only analyze the behavior of the flow for t close to zero, since any metric in the flow may be chosen to be the base metric. This point is used several times in section 2.

References

[1] R. Arnowitt, S. Deser and C. Misner, "Coordinate Invariance and Energy Expressions in General Relativity," *Phys. Rev.* **122** (1961) 997-1006.

[2] R. Bartnik, "The Mass of an Asymptotically Flat Manifold," *Comm. Pure Appl. Math.* **39** (1986) 661-693.

[3] R. Bartnik, "New Definition of Quasi-Local Mass," *Phys. Rev. Lett.* **62** (1989) 2346.

[4] R. Bartnik, "Quasi-Spherical Metrics and Prescribed Scalar Curvature," *J. Diff. Geom.* **37** (1993) 31-71.

[5] H. L. Bray, "The Penrose Inequality in General Relativity and Volume Comparison Theorems Involving Scalar Curvature," thesis, Stanford University, 1997.

[6] H. L. Bray, "Proof of the Riemannian Penrose Conjecture Using the Positive Mass Theorem," (in preparation).

[7] H. L. Bray, F. Finster, "Curvature Estimates and the Positive Mass Theorem," (submitted).

[8] Bunting, Masood-ul-Alam, "Non-Existence of Multiple Black Holes in Asymptotically Euclidean Static Vacuum Space-Time," *Gen. Rel. and Grav.*, Vol. 19, No. 2, 1987.

[9] D. Christodoulou, "Examples of Naked Singularity Formation in the Gravitational Collapse of a Scalar Field," *Ann. of Math.* **140** (1994) 607-653.

[10] D. Christodoulou and S.-T. Yau, "Some Remarks on the Quasi-Local Mass," *Contemporary Mathematics* **71** (1988) 9-14.

[11] R. Geroch, "Energy Extraction," *Ann. New York Acad. Sci.* **224** (1973) 108-17.

[12] G. Gibbons, "Collapsing Shells and the Isoperimetric Inequality for Black Holes," *Univ. of Cambridge*, preprint, 1997.

[13] S. W. Hawking, "Gravitational Radiation in an Expanding Universe," *J. Math. Phys.* **9** (1968) 598-604.

[14] S. Hawking, *Phys. Rev. Lett.* **26**, 1344 (1971).

[15] S. W. Hawking, "Black Holes in General Relativity," *Comm. Math. Phys.*, **25** (1972) 152-166.

[16] S. W. Hawking and G. F. R. Ellis, *The Large-Scale Structure of Space-Time*, Cambridge University Press, Cambridge, 1973.

[17] S. W. Hawking and R. Penrose, "The Singularities of Gravitational Collapse and Cosmology," *Proc. Roy. Soc. A* **314** (1970) 529-548.

[18] M. Herzlich, "A Penrose-like Inequality for the Mass of Riemannian Asymptotically Flat Manifolds," to appear in *Comm. Math. Phys.*

[19] G. Huisken and T. Ilmanen, "The Inverse Mean Curvature Flow and the Riemannian Penrose Inequality," *J. Diff. Geom* (to appear).

[20] G. Huisken and S.T. Yau, "Definition of Center of Mass for Isolated Physical Systems and Unique Foliations by Stable Spheres with Constant Mean Curvature," *Invent. Math.* **124** (1996) 281-311.

[21] W. Israel, *Phys. Rev.* **164**, 1776 (1967); *Comm. Math.* Phys. **8**, 245 (1968).

[22] P.S. Jang, "On the Positive Energy Conjecture," *J. Math. Phys.* **17** (1976) 141-145.

[23] P.S. Jang, "On the Positivity of Energy in General Relativity," *J. Math. Phys.* **19** (1978) 1152-1155.

[24] P.S. Jang, "Note on Cosmic Censorship," *Phys. Rev. Lett. D* **20** (1979) 834-838.

[25] P.S. Jang, "On the Positivity of the Mass for Black Hole Space-Times," *Comm. Math. Phys.* **69** (1979) 257-266.

[26] P.S. Jang and R. M. Wald, "The Positive Energy Conjecture and the Cosmic Censor Hypothesis," *J. Math. Phys.* **18** (1977) 41-44.

[27] R. Penrose, "Naked Singularities," *Ann. New York Acad. Sci.* **224** (1973) 125-134.

[28] D. Robinson, *Phys. Rev. Lett.* **34**, 905 (1975).

[29] R. Schoen, "Variational Theory for the Total Scalar Curvature Functional for Riemannian Metrics and Related Topics," *Topics in Calculus of Variations (M. Giaquinta, ed.) Lecture Notes in Math.*, **1365**, 120-154, Springer, Berlin, 1987.

[30] R. Schoen and S.-T. Yau, "Incompressible Minimal Surfaces, Three-Dimensional Manifolds with Nonnegative Scalar Curvature, and the Positive Mass Conjecture in General Relativity," *Proc. Nat. Acad. Sci.* **75** (1978), no. 6, 2567.

[31] R. Schoen and S.-T. Yau, "On the Proof of the Positive Mass Conjecture in General Relativity," *Comm. Math. Phys.* **65** (1979) 45-76.

[32] R. Schoen and S.-T. Yau, "Positivity of the Total Mass of a General Space-Time," *Phys. Rev. Lett.* **43** (1979) 1457-1459.

[33] R. Schoen and S.-T. Yau, "Proof of the Positive Mass Theorem II," *Comm. Math. Phys.* **79** (1981) 231-260.

[34] R. Schoen and S.-T. Yau, "The Energy and the Linear Momentum of Space-Times in General Relativity," *Comm. Math. Phys.* **79** (1981) 47-51.

[35] R. Schoen and S.-T. Yau, "The Existence of a Black Hole due to Condensation of Matter," *Comm. Math. Phys.* **90** (1983) 575-579.

[36] K. P. Tod. *Class. Quant. Grav.* **9** (1992) 1581-1591.

[37] E. Witten, "A New Proof of the Positive Energy Theorem," *Comm. Math. Phys.* **80** (1981) 381-402.

Current Developments in Mathematics, 1999

Recent Progress in Sphere Packing

J. H. Conway, C. Goodman-Strauss, and N. J. A. Sloane

Most of this article will be about comparatively minor progress in the sphere packing problem in higher dimensions, but it is a pleasure to record that at last the original 3-dimensional density problem has been definitively solved. The attempted proof by Hsiang [**Hs93**] unfortunately turned out to be incomplete.

The Hales Proof of the Kepler Conjecture for 3-dimensional Sphere Packings

The problem of finding the greatest density of any packing of equal spheres in n-dimensional Euclidean space is an old and important one, which has applications in geometry, number theory, and information theory. It is also difficult, as is shown by the fact that although the 3-dimensional case was raised by Kepler in 1613, it has only recently been solved, by Thomas Hales in 1998, with an important contribution from S. Ferguson. I shall say only a few words about the argument, which is long and complicated in detail.

Hales starts from the familiar Delaunay tessellation, which has one cell for each point of space whose distance from the sphere centers is a local maximum, for which Sloane and I have introduced the snappy term "deep hole". Then the Delaunay cell that corresponds to a deep hole is the convex hull of the sphere centers nearest to that hole, and it is well known that these cells constitute a polyhedral decomposition of space.

In the standard face-centered cubic lattice packing, the Delaunay cells are alternately regular tetrahedra T and octahedra O, so that its density is

$$[vol(T') + vol(O')]/[vol(T) + vol(O)]$$

where T' and O' are the parts of T and O covered by the spheres. However, for a packing in which the sphere centers are in general position the Delaunay cells will be simplicial, and so it suffices to establish the density bound for packings with simplicial Delaunay cells.

C. A. Rogers long ago showed that $vol(T')/vol(T)$ was an upper bound for the density of any packing—this is now known as "the Rogers bound". It would be attained for a hypothetical packing whose Delaunay cells were all regular tetrahedra, so we may say that the difficulty in the 3-dimensional sphere packing problem is to show that the Rogers bound must inevitably be worsened to the extent that it is in the face-centered cubic packing by the presence of the octahedral cells.

In the first paper of his series, Hales defined a function called the "score" of a star of Delaunay cells, in terms of another function called the "compression" of a cell. The compression of the Delaunay simplex S is

$$vol(S') - vol(S).vol(O')/vol(O)$$

where S' is the part of S contained in the spheres, and O and O' are as above. In other words, the compression measures how much better S is covered than is the regular octahedral cell O of the face-centered cubic packing.

A tetrahedron S is naturally divided into four parts (the Voronoi cells of its vertices), the nth part S_n consisting of those points that are closer to the nth vertex than any other. In the case that S contains its circumcenter, its score at the nth vertex is defined to be its compression, if the circumradius is at most 1.41, and

$$4.solid(n)/3 - 4.vol(S_n).vol(O')/vol(O)$$

otherwise, where $solid(n)$ denotes the solid angle at that vertex. When S does not contain its circumcenter, the score is defined by continuing the analytic function corresponding to this expression. The score (measured in "points") of the star of Delaunay tetrahedra for a given vertex is the sum of these scores over all the tetrahedra at that vertex. These definitions ensure that the average of the score over all of space is the average of the compression, and reduce the problem to proving that the average score is at most 8 points.

Two sphere centers are called "close neighbors" if they are distant at most 2.51 from each other, and a Delaunay cell is called a quasi-regular tetrahedron if any two of its four vertices are close neighbors. In his first paper, Hales shows that the score of a star composed entirely of quasi-regular tetrahedra is indeed at most 8 points. The remaining papers of the sequence are addressed to the much harder task of extending this result to all the other configurations that might arise.

This involves a combinatorial classification of the possibilities, which is in itself very long, accompanied by an analytical proof of the appropriate inequality for each of thousands of cases. Both parts involved heavy use of machine computation, and careful selection of various parameters. Hales remarks for instance that

"The constant 2.51 was determined experimentally to have a number of desirable properties", and similar experimental determinations recur repeatedly throughout the paper. Several of the initial decisions had to be modified in the light of later calculations.

Of course a proof of this nature is not at all easy to read! But Hales and Ferguson have taken great care to make the entire proof accessible to readers who wish to check any detail. In particular they have started a practice that should serve as a model for future machine-dependent proofs, by keeping detailed logs of all their interactions with the machine, so that a potential auditor can discover that on such and such a day the following cases were handled, of which the last had to be split into two subcases for which the inequalities became ... . He or she can then rerun the programs which were used to verify the truth of these inequalities.

Great care was also taken with these proofs. Suppose, for instance, that the inequality $f(x) < g(x)$ has to be proved for all x in a certain interval. Then typically the machine will automatically find a dissection of this interval into a number of subintervals, and find upper bounds for the derivatives of f and g at the endpoints of these which yield linear bounds for them that still satisfy the inequality. [For higher-dimensional intervals this would involve linear programming.] All calculations of such bounds are done using interval arithmetic, which prevents errors that might otherwise arise from rounding the numbers.

Progress in other dimensions: The problem of "best" packings.

In other dimensions, our progress has been limited, and much of it only conjectural. The question we should like to answer is

"What are all the best sphere packings in a given dimension?"

Unfortunately, it is not even clear just what this question means! Despite this, the paper **[CS95]** of Conway and Sloane gives a conjectural answer to it in each dimension up to 10. We shall roughly follow the discussion in that paper, marching onward from dimension to dimension, discussing both the question and its answers as we go. As the reader may well suspect, packings of R^0 are not particularly intriguing, and so we begin with:

What are the best sphere packings in $\mathbf{R}^1$?

We can regard a sphere packing by spheres of radius r as a collection of points each pair of which is at least r apart. No matter what the best definition of "best" may be, the best packing of spheres of radius r in $\mathbf{R}^1$ is surely the lattice of points $2kr$:

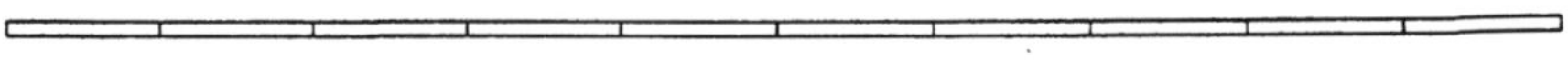

And clearly, any packing that can be compacted further should not be called "best":

Note, perhaps a bit pedantically, that the spheres in a best packing in $\mathbf{R}^1$ are centered on the points of a lattice; we can rescale this to be the root lattice A_1, that is, the lattice generated by the vector $(1, -1)$ in $\mathbf{R}^2$. Root lattices will play a large role in our discussion of best packings; and since the root lattices are canonically scaled so that the minimum distance between points is $\sqrt{2}$, we will adopt the convention that all of our sphere packings are packings of spheres of radius $\sqrt{2}/2$.

What are the best sphere packings in $\mathbf{R}^2$?

Whatever "best" may mean, the best packing of the plane should of course be the hexagonal packing (Figure 1), with circles centered on the points of the root lattice A_2, the lattice generated by the vectors, say, $(1, -1, 0)$ and $(0, 1, -1)$:

Clearly "having the highest possible density" is a major component of any good meaning of "best", and Fejes Tóth has given an elegant proof that no packing in R^2 has density exceeding that of the hexagonal lattice A_2. However, if we allow this to be the entire definition we get some rather silly "best packings" since density is only defined by a limiting process.

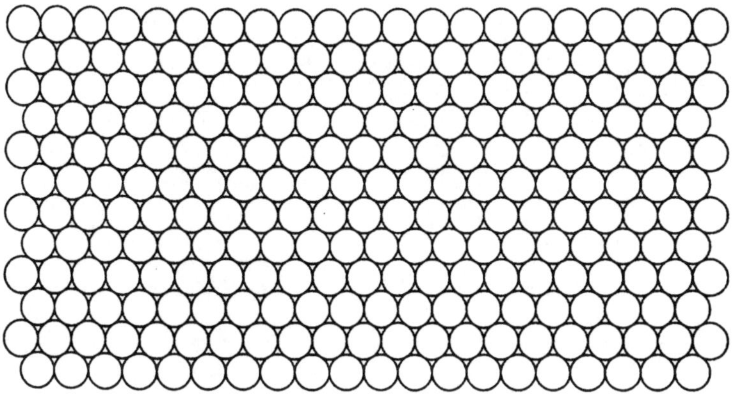

FIGURE 1. The A_2 lattice packing, Π_2

For example, beginning with this hexagonal packing, remove any finite number of circles. The depleted packing still has the same density, as does the "even worse" one in which we remove an entire infinite row of circles. In fact, we can even have a packing of the optimal density in which no circles are touching! Figure 2 below is derived from the hexagonal lattice by increasing all distances from the origin by a fixed multiple of their arctangents, an operation that does not change the overall density of the packing.

Perhaps we would do better with Tóth's notion of solidity: a packing is "solid" if there is only one way to replace any finite collection of spheres we have removed. Here is a little theorem:

Theorem 1. *If two packings in* $\mathbf{R}^n$ *have different densities, then the looser cannot be solid.*

The proof is quite simple. Consider the looser packing. If we carve out a vast but still finite region, we will be free to put all the spheres back in the denser configuration, with plenty of room around the sides. For example, in Figure 3, we have replaced 143 circles in the relatively loose square packing with 143 circles from a much denser hexagonal packing.

Solidity, then, implies having the highest density, and is a good candidate for the meaning of "best". But unfortunately there seem to be some not-so-good examples of solid packings. For example, consider the packing in Figure 4. The "upper half" of a hexagonal packing has been shifted slightly; the shaded circles show the fault line. Though this packing is clearly not "best", it is almost certainly solid.

Worse still it is extremely difficult to establish the solidity of any particular packing. The recent work of Hales, described above, does *not* suffice to establish the solidity of any Barlow packing.

But it is even difficult to show that there *is* a solid packing in a given space. It is possible, and quite plausible in higher dimensions, that solid packings simply don't exist.

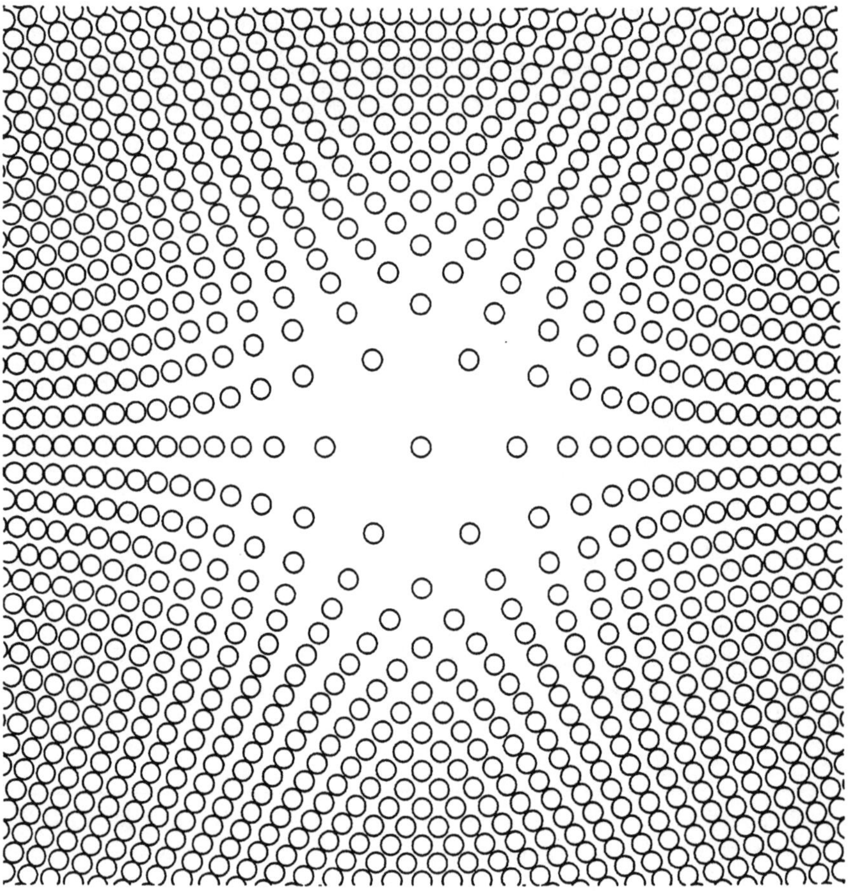

FIGURE 2. A bad but maximally dense packing in $\mathbf{R}^2$

We might define "best" as something like "having maximum density over finite regions". But this definition seems to fail. Beyond the completely compacted packing in $\mathbf{R}^1$, every sphere packing in $\mathbf{R}^n$ has plenty of empty space; over just which finite regions should we measure this density?

So just how should we define "best"? Conway and Sloane were unable to find a satisfactory definition and eventually gave up on this aspect of the problem. This does not force us to give up on the real and apparently meaningful problem of finding out just what the "best" packings actually are. Since in any case, we can't reasonably expect to find more than a conjectural answer to the question, the fact that we can't assign a precise meaning to it is not as distressing as it might have been!

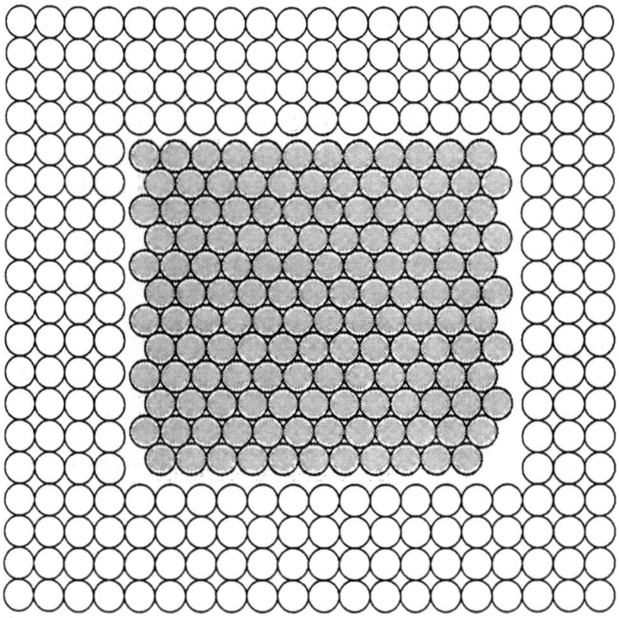

FIGURE 3. Theorem 1

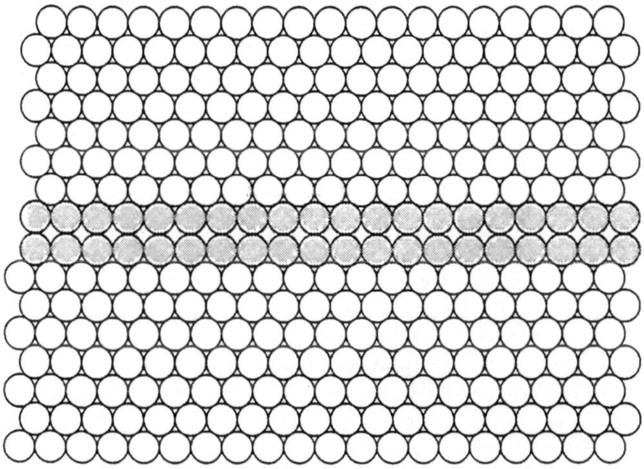

FIGURE 4. "Solid" $\neq$ "best"

What are the "best" sphere packings in $\mathbf{R}^3$?

We have already commented on one of the most spectacular developments in recent years, Thomas Hales' settling of the so-called Kepler Conjecture, that no packing in $\mathbf{R}^3$ has greater density than that based on the face-centered cubic lattice $A_3 \cong D_3$, *viz.*, the points of the form (i, j, k) with $i + j + k$ even.

As it happens, the face-centered cubic lattice packing is far from being uniquely the densest; it is only the most symmetrical of an uncountable family of equally good "Barlow packings", named for the great English crystallographer who studied some of them in 1883 [**Ba83**].

It is conceivable that there are other packings in $\mathbf{R}^3$ that deserve the name "best", for Hales showed only that the density of the Barlow packings cannot be surpassed. Nonetheless, despite the lack of a sound definition of "best", it is quite reasonable to suggest that the Barlow packings are indeed all the best packings in $\mathbf{R}^3$.

To make a Barlow packing of your own, begin with a sheet of spheres arranged into a two-dimensional hexagonal packing. There are precisely two ways to stack another such sheet neatly on top of the first: we can fit the next layer of spheres above either of two sets of holes, as shown in Figure 5. Repeating the process, we face two choices at each layer and altogether can produce an uncountably infinite family of packings as promised, all with exactly the same density.

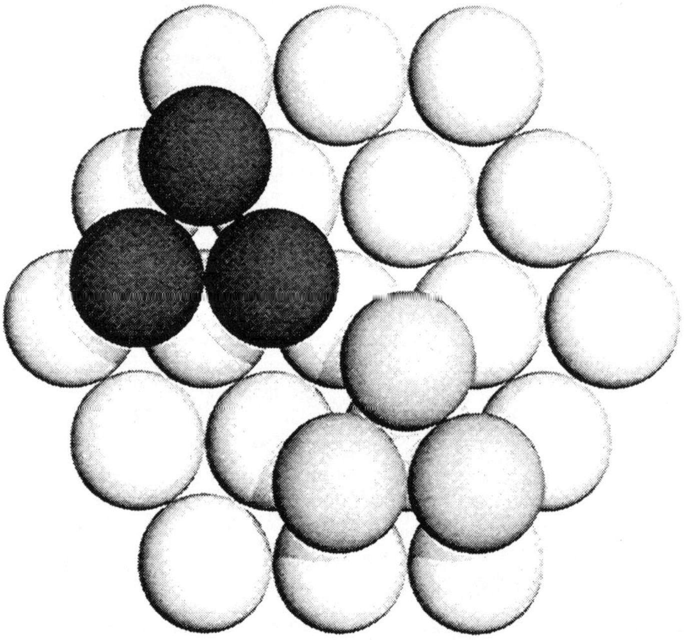

FIGURE 5. Stacking hexagonal lattice packings

Only two of these are "uniform", however, in the sense that there is a symmetry of the packing taking any one sphere to any other. To see this, look down upon a

Barlow packing from above. All the centers of the spheres in any given layer lie in one of three sets of points, colored white [0], gray [1] and black [2] in figure 6. In fact, these points form the unique three-coloring of the hexagonal lattice A_2. To put it another way, the three sets of points represent three cosets of the unique sub-lattice of index 3 in A_2; this sub-lattice is of course isomorphic to A_2 itself.

FIGURE 6. Three cosets in A_2.

We call the possible projections of the layers *placements*, and label them by the "colors" [0], [1] and [2]. Now there is only one restriction on the possible sequence of placements of the layers in the Barlow packing, namely that we never use the same color for two adjacent layers. So the Barlow packings are exactly in correspondence with doubly infinite paths in this *placement graph*, 3_2 (we will use v_k to denote a graph with v vertices each of valence k):

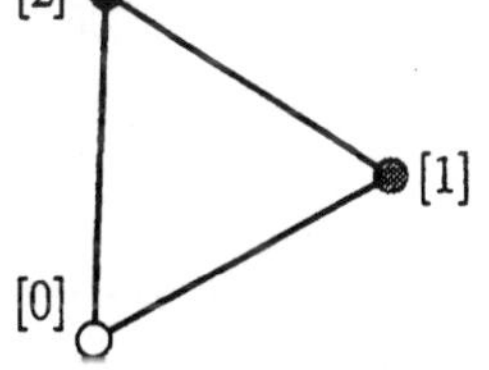

Alternatively, the Barlow packings are in precise correspondence with the 3-colorings of A_1.

Now any *uniform* packing exactly corresponds to a path whose vertices are equivalent under any symmetries of this graph. Clearly, up to labeling, there are only two such paths: ... abababababa ... and ... abcabcabcabc ... where a, b, c is some permutation of the colors $[0], [1], [2]$. Small neighborhoods of these two packings are shown in Figure 7.

The packing in which the layers cyclically permute the colors is the more special; this is the face-centered lattice packing with which we began, and it is the only *lattice* packing among the Barlow packings.

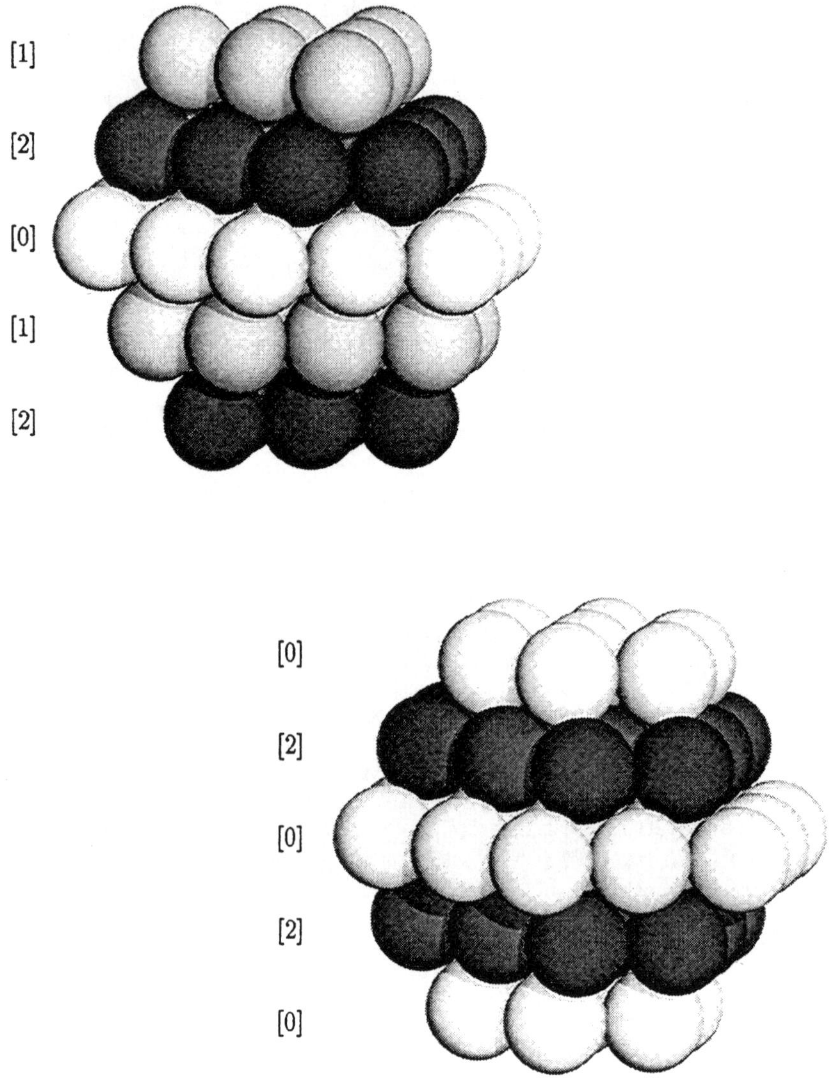

FIGURE 7. Portions of the two uniform Barlow packings: the face-centered cubic lattice packing and the hexagonal close packing

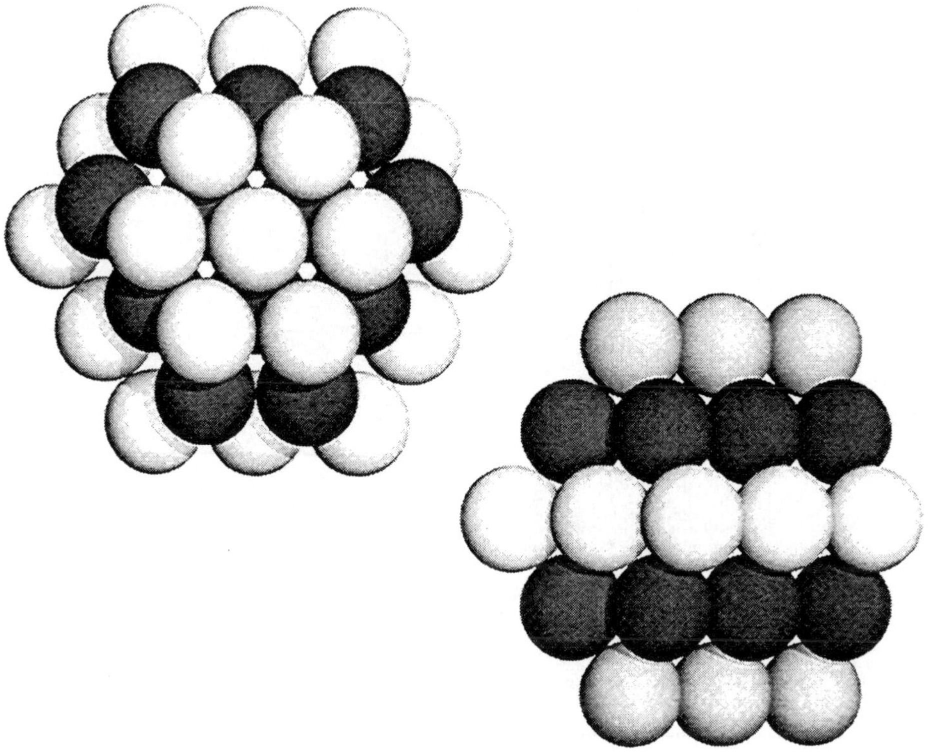

FIGURE 8. No Barlow packing completely occludes all light ...

It is quite amusing to note that the other uniform packing, the so-called hexagonal close-packing, in which the layers occur in just two alternating colors, is the "least occluding" of all the Barlow packings.

As Figure 8 suggests, no Barlow packing completely occludes all light (part of a generic packing, is on the lower right; the close-packed hexagonal packing, seen from a different vantage is on the upper left.) Equivalently, one can always drive rather thin, infinitely long stakes in one of three "horizontal" directions, without touching any of the spheres in the packing. In the face-centered cubic lattice, there is enough additional symmetry that one can drive thin stakes in three more directions. But the hexagonal close packing is the only Barlow packing in which there is enough room for really thick stakes, more than 1/7th the width of the spheres themselves. These stakes are "vertical" and run through the holes above the third, unused coset in A_2, as seen in Figure 9.

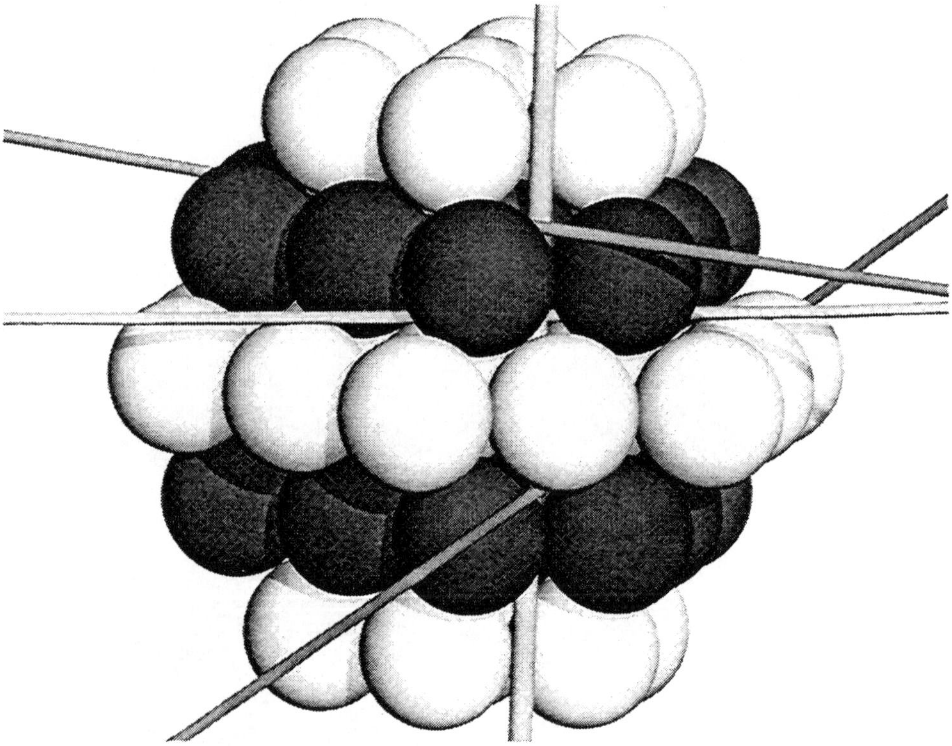

FIGURE 9. ... but the hexagonal close packing is the least occluding of all

Crystal Balls. The *first crystal ball* of a sphere in a packing consists of that sphere and those touching it, and inductively, the $(n+1)st$ *crystal ball* consists of the nth crystal ball and all the spheres that touch it.

In all of the Barlow packings, every sphere has exactly 12 neighbors, so the first crystal ball of any sphere has size 13. However, the *second* crystal ball can have size 55, 56, or 57. Consider five adjacent layers labeled a, b, c, d, e, where each color is one of [0], [1] or [2]. Then the second crystal ball has size 55 plus the number of "Yes" answers to the two questions:

$$\text{Is } a = c? \qquad \text{Is } c = e?$$

The case when $a \neq c$ and $c = e$, so that there are 56 spheres in the second crystal ball, is shown in Figure 10.

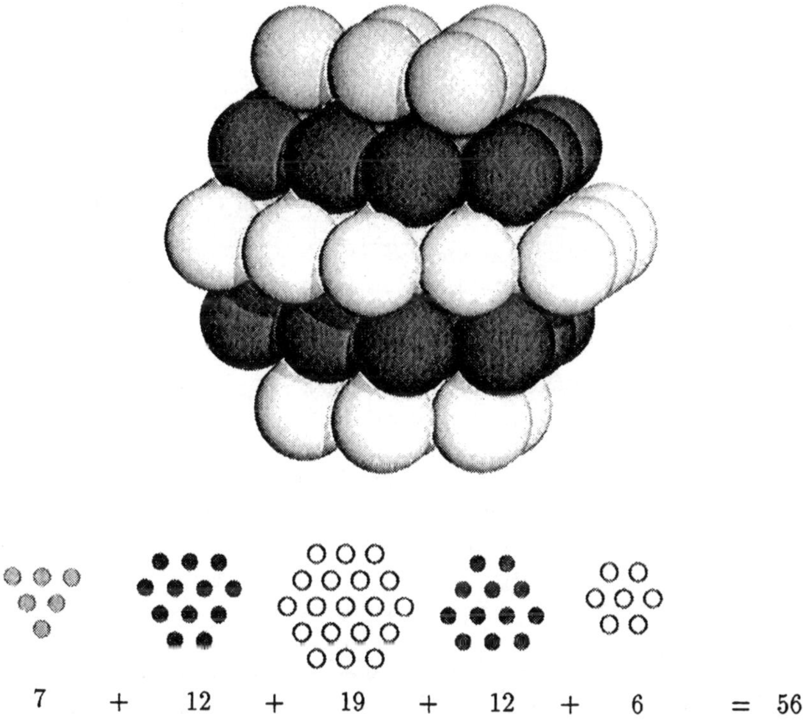

FIGURE 10. A second crystal ball has 56 spheres when $a \neq c$ and $c = e$

There are just four packings that are determined by the set of sizes of their second crystal balls:

If no crystal ball has size 56, then the answers to alternate questions from the sequence

$$\ldots \qquad \text{Is } a = c? \qquad \text{Is } b = d? \qquad \text{Is } c = e?$$

must always agree and we obtain one of three packings:

Packing	Answers	Coloring	Sizes
Λ_3^1	... NNNN ...	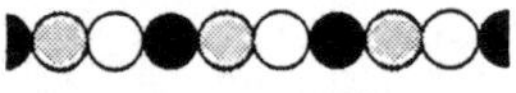	55
Λ_3^2	... YYYY ...		57
Λ_3^3	... YNYN ...		55,57

We have already seen Λ_3^1 and Λ_3^2; these are the two uniform packings, *viz.,* the face centered cubic lattice packing and the hexagonal close packing.

If all the crystal balls have size 56, then alternate answers must always be different and we have a fourth packing

Λ_3^4 ... YYNNYYNNYY ...  56

There are uncountably many Barlow packings with some second crystal balls of size 56 and some with size 55 or 57 or both, so these four are the only Barlow packings determined by the sizes of their second crystal balls.

Since all of the second crystal balls in the face-centered cubic lattice packing contain 55 balls, and all of those in the hexagonal close packing contain 57 balls, the average number of spheres in the second crystal ball in any large region of a Barlow packing is therefore between that of the two uniform Barlow packings. And so it is natural to suspect that the number of spheres in the nth crystal ball of any sphere in any Barlow packing is bounded above by the size of an nth crystal ball in the hexagonal close packing, and below by the size of the nth crystal ball in the face-centered cubic lattice packing.

Conway and Sloane verefied this by showing that the number of spheres in any nth crystal ball in the face centered-cubic lattice packing is $(5\Delta n^4 + \Delta n^2)/6$ where $\Delta n^k = (n+1)^k - n^k$; similarly the number of spheres in any nth crystal ball in the hexagonal close packing is $[7\Delta n^4/8]$. And indeed the number of spheres in any nth crystal ball in any Barlow packing is bounded below by the former and above by the latter (a conjecture that was only later made by O'Keefe).

The Hypothetical Answers

The Barlow packings generalize quite nicely to a conjectured description of *all* the best packings in low dimensions:

A n-dimensional packing P_n *fibers* over an m-dimensional packing P_m if P_n can be decomposed into sets (e.g. layers) of points lying in parallel m-dimensional spaces, each one of which is a packing of type P_m. In particular, then, the Barlow packings all fiber over the hexagonal packing in R^2.

Hypothesis n : *The best packings in $\mathbf{R}^n$ fiber over one of the best packings in $\mathbf{R}^m$, where m is the largest power of 2 less than n.*

Conway and Sloane believe (but cannot prove) that Hypothesis n is true for $2 \leq n \leq 8$. Hypothesis 9 requires some adjustments, and the Hypotheses are irredeemably false for $n \geq 10$.

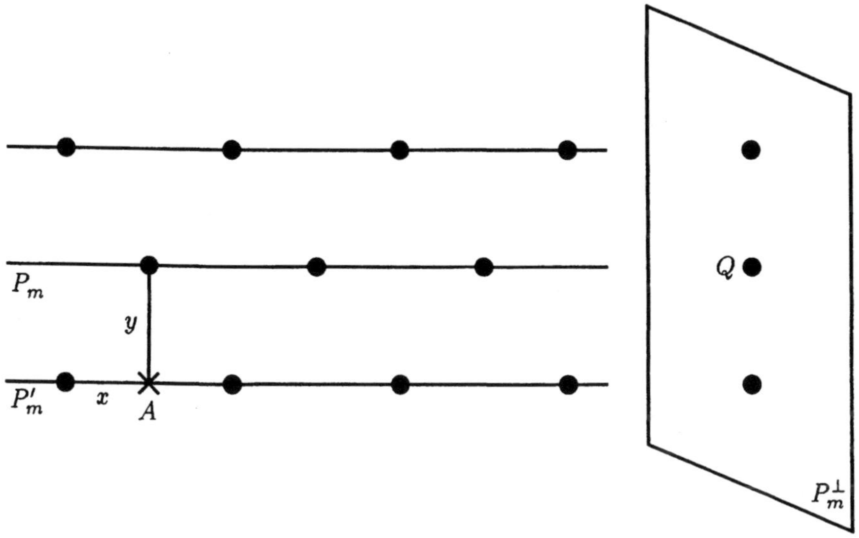

FIGURE 11

Let us fix some ideas: Suppose P_n is an n-dimensional packing that fibers over an m-dimensional packing P_m. Let A be the projection onto the space of a layer P_m of a point in another layer P'_m. Let x be the distance of A from the closest point of P_m and let y be the separation between the layers P_m and P'_m as illustrated in Figure 11. Then x is at most the covering radius R of P_m and $x^2 + y^2 \geq 2$ (because our spheres are of radius $1/\sqrt{2}$, so y is at least $\sqrt{2 - R^2}$).

Let Q be the projection of P_n onto the space $P_m^\perp$. Surrounding each point of Q by a sphere of diameter $\sqrt{2 - R^2}$, we obtain a sphere packing in $P_m^\perp$. If P_n is to be "best", plainly the packing in $P_m^\perp$ must be "best" (The hypothesis asserts that all best packings can be obtained this way taking m to be the greatest power of 2 less than n.) This can only occur if for adjacent layers P_m and P'_m, every point of P_m lies above a deep hole of P'_m.

Before marching off at the rapid rate the Hypotheses will allow us, let us turn back and see Hypotheses 2 and 3 applied to produce the best packings in R^2 and R^3.

Proposition 2 (using Hypothesis 2) *The unique "best" packing in* $\mathbf{R}^2$ *is the triangular packing* A_2.

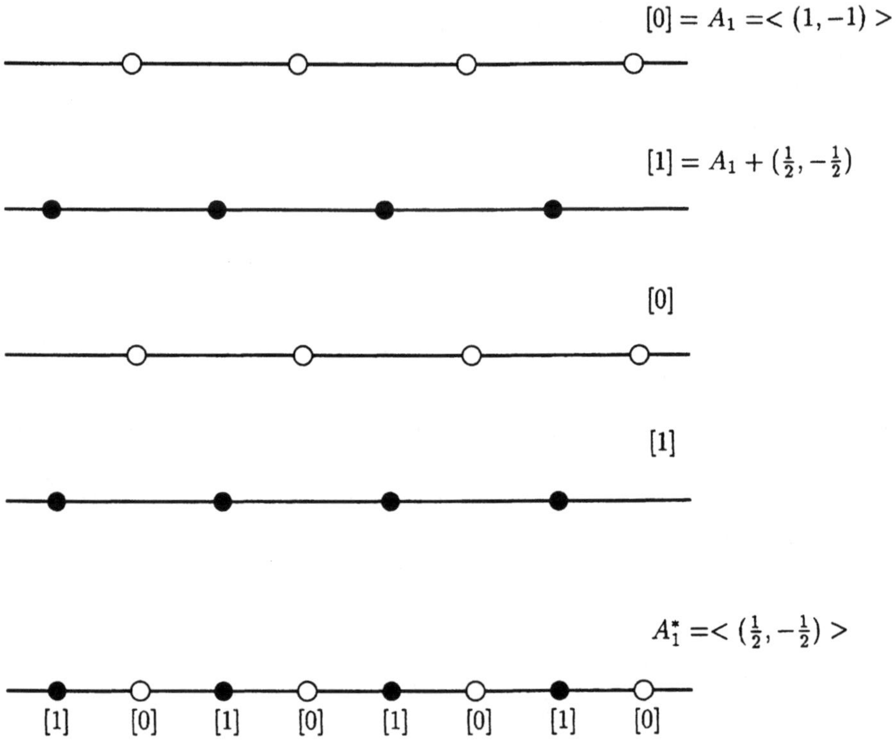

FIGURE 12. Applying Hypothesis 2 ...

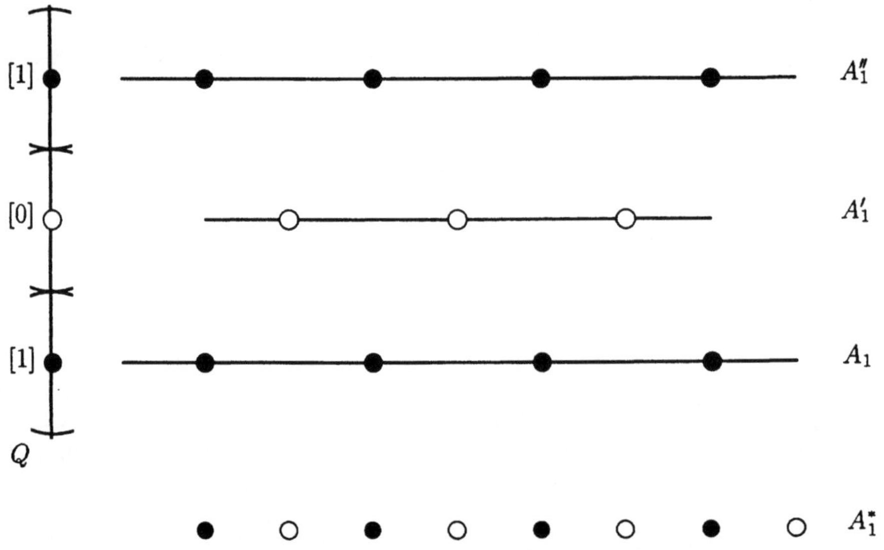

FIGURE 13. ... to prove Proposition 2

Proof. Hypothesis 2 tells us that any best packing in $\mathbf{R}^2$ fibers over A_1, the unique best packing in $\mathbf{R}^1$. The dual lattice A_1^* is the union of two cosets $[0] = A_1$ and $[1] = A_1 + (\frac{1}{2}, -\frac{1}{2})$. Each of the cosets consists of the deep holes of the other. The covering radius of A_1 is $1/\sqrt{2}$. The distance y between layers must be at least

$\sqrt{2-1/2} = \sqrt{3/2}$; this can be achieved only for the packing shown in Figure 13, in which alternate layers are the cosets [0] and [1] of A_1 in A_1^*. The packing Q is a rescaled A_1, in which the points are colored alternately [0] and [1]. Of course, this is a description of the root lattice A_2. □

As with the Barlow lattices, we can describe the *placements* of the layers of the A_2 lattice packing. The allowed sequences of possible placements corresponds to paths in the *placement graph*, in this case, merely the graph 2_1:

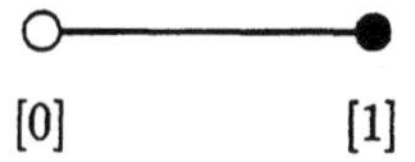

More generally, the layers in a "best" packing can only have a certain set of projections onto the starting layer; these are the placements. Two placements are joined by an edge in the placement graph just if they can correspond to layers at the minimal distance $\sqrt{2-R^2}$, that is, just if each point of either projects into a deep hole of the other.

Fibering over $\mathbf{A_2}$ Several times we will be fibering over A_2. The lattice A_2^* dual to A_2 consists of the three cosets

$$[0] = A_2, \quad [1] = A_2 + \frac{1}{3}(2,-1,-1), \quad [2] = A_2 + \frac{1}{3}(-2,1,1)$$

of A_2. The set of deep holes in any one of these cosets is the union of the other two cosets, and the covering radius is $\sqrt{\frac{2}{3}}$. Each of the cosets [0], [1], [2] and their union A_2^* is a version of the hexagonal lattice. This is, of course, precisely the illustration shown earlier in Figure 6.

Proposition 3 (using Hypotheses 2 and 3) *The "best" packings in* $\mathbf{R}^3$ *are the Barlow packings.*

We already have given a rather informal description of the Barlow packings as something like a fibering. Let's go through the process more carefully:

Proof. Let Π_3 be a best packing in R^3. Hypotheses 2 and 3 tell us that each layer in any Π_3 is a copy of A_2 and that the separation between any two layers is at least $\sqrt{2-2/3} = \sqrt{4/3}$. Now suppose adjacent layers are always at this distance, and choose any particular layer to be $[0] = A_2$. Then the next layer must be a copy of A_2 contained in $[1] \cup [2]$. However, since squared distances in A_2 are even integers, the next layer cannot contain points from both [1] and [2] (the squared distances between which have the form $(2n+2/3)$ and so must be all of [1] or all of [2], say [1]. The layer after this must be [0] or [2], etc. This is of course the way we described the Barlow lattices earlier. □

Now we use the hypotheses to make short work of the next several dimensions:

What are all the "best" packings in $\mathbf{R}^4$?

Proposition 4 (using Hypotheses 2 and 4) *The unique best packing in* $\mathbf{R}^4$ *is the root lattice* D_4

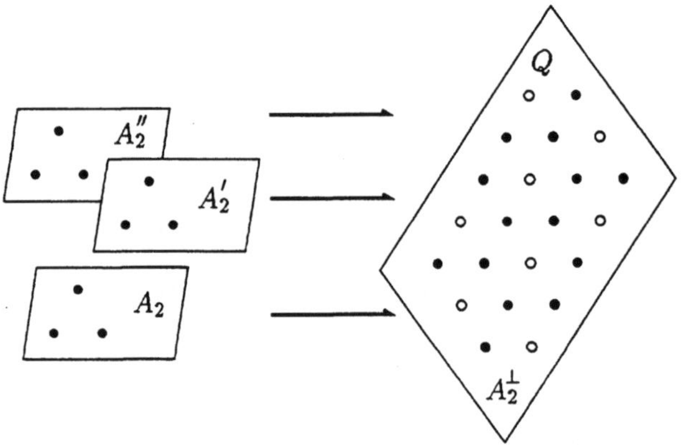

FIGURE 14. In the proof of Proposition 4, both fiber and quotient are two-dimensional

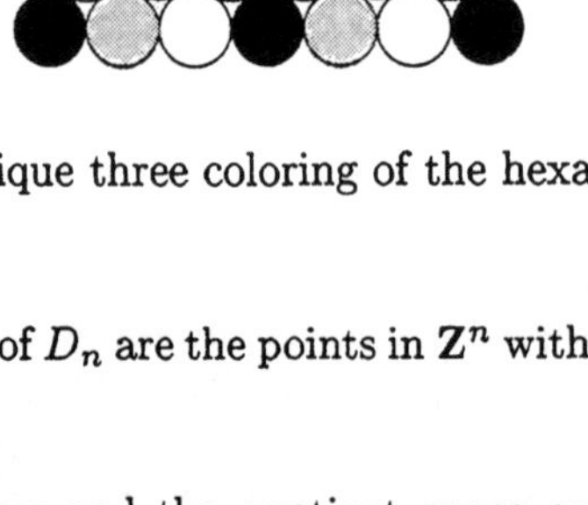

FIGURE 15. The unique three coloring of the hexagonal lattice packing

Recall that the points of D_n are the points in $\mathbf{Z}^n$ with the sum of the coordinates even.

Proof. Here both the layers and the quotient space are two-dimensional (Figure 14, and any best packing must come from a three-coloring of the two-dimensional packing Q, a rescaled version of A_2. However, this three-coloring is unique (Fig. 15) and is completely determined by the colors of any three mutually adjacent circles. The resulting lattice is D_4. □

The dual lattice D_4^* consists of four cosets

$$[0] = D_4, \quad [1] = D_4 + \frac{1}{2}(1,1,1,1), \quad [2] = D_4 + (0,0,0,1)$$

$$[3] = D_4 + \frac{1}{2}(1,1,1,-1)$$

and the set of deep holes in any one of these cosets is the union of the other three. The covering radius is 1.

What are all the "best" packings in $\mathbf{R}^5$?

Proposition 5 (using Hypotheses 2, 4 and 5) *The best packings in $\mathbf{R}^5$ are parametrized by paths along the edges of a tetrahedron. Just four of them are uniform packings, and just one of these is a lattice packing.*

Proof. The usual argument shows that any two layers must be separated by at least $\sqrt{2-1} = 1$, and that if two layers D_4 and D_4' are separated by just this distance, the points of D_4' must lie above the deep holes of D_4. In other words, D_4' projects into the union of the cosets [1], [2] and [3]. As before, this image cannot contain points from more than one of these cosets and so must be a single coset [1], [2] or [3]. So this time, we have four possible placements for each layer, and once again the only restriction is that we may not use the same placement twice in succession. The placement graph is thus a tetrahedron, 4_3:

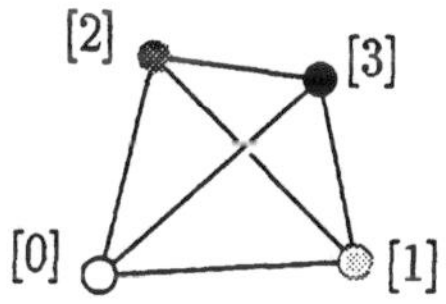

Any "best" five-dimensional packing is given by a sequence of placements

$$\ldots abcde \ldots$$

which corresponds to a doubly infinite path along the placement graph.

For a uniform packing, there are several cases. First, if only two placements are used, say [0] and [1], we have the root lattice D_5 corresponding to the coloring

Λ_1^5 

The corresponding path, of course, winds back and forth across a single edge of the tetrahedron (shown at left in Figure 16). D_5 has quite a bit of symmetry and can be decomposed into layers in several such ways.

In any other uniform packing any three adjacent latters must be distinct. Moreover, the division into layers is characteristic: adjacent spheres A and B are in the same layer just if there is a third sphere C touching B antipodally to A.

We now consider the spheres within the second crystal ball of a given sphere, in layer c, say, among the five consecutive layers a, b, c, d, e, and we can ask the two questions

$$\text{Is } a = d? \qquad \text{Is } b = e?$$

The set of answers must be the same for every sphere and determines the packing.

Λ_2^5 NN

Λ_3^5 NN

Λ_4^5 NN

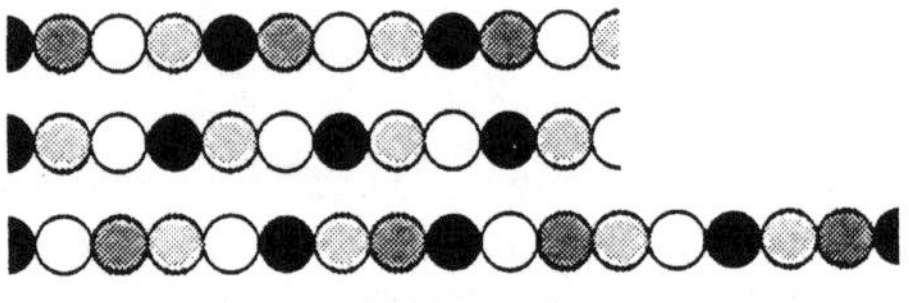

In Figure 16 we see the paths in 4_3 that give rise to these four uniform best packings in $\mathbf{R}^5$.

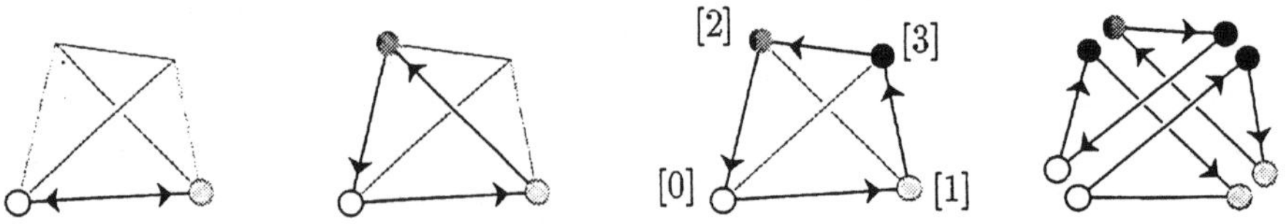

FIGURE 16. Paths in 4_3 giving rise to uniform best packings in $\mathbf{R}^5$

Thus there are four uniform packings, first found by Leech [**Le67**]. It turns out that Λ_5^n consists of n translates of a lattice packing ($n = 1, 2, 3, 4$) □.

What are all the "best" packings in $\mathbf{R}^6$?

Proposition 6 (using Hypotheses 2, 4 and 6) *The best packings in $\mathbf{R}^6$ are parametrized by 4-colorings of the A_2 lattice. Just four of them are uniform.*

Proof. The main assertion goes as before.

Let us consider the the colors of six neighbors of a circle colored d in A_2. There are essentially just four possibilities for the colors:

abcabc
abacbc
ababab
ababac

For a uniform packing, every circle in the A_2 quotient must be surrounded in the same way as every other. It happens that each of the four types gives a unique uniform coloring, as shown in Figure 17, the first being the root lattice $\Lambda_6 = E_6$. Once again, the packing Λ_6^n consists of n translates of a lattice ($n = 1, 2, 3, 4$). □

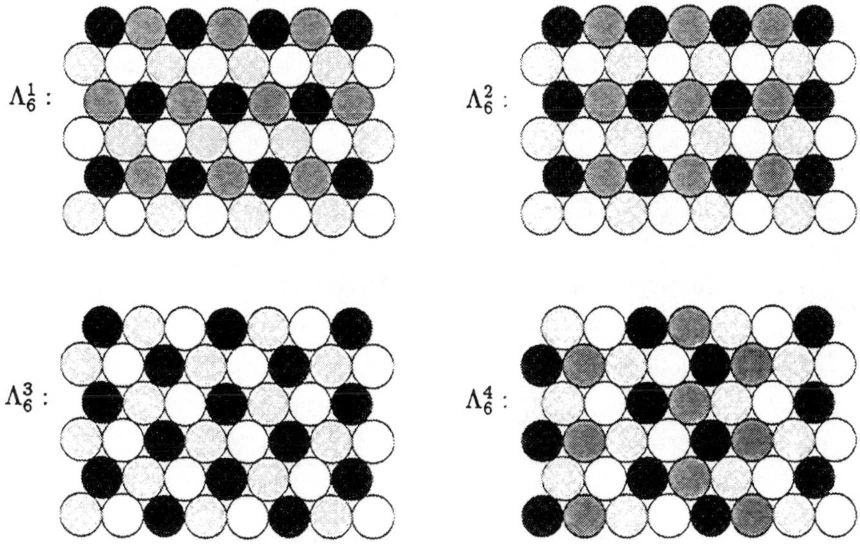

FIGURE 17. The four colorings that give the uniform best packings in $\mathbf{R}^6$

What are all the "best" packings in $\mathbf{R}^7$?

Proposition 7 (using Hypotheses 2, 3, 4 and 5) *The best packings in $\mathbf{R}^7$ fiber over D_5 and can be parametrized by choosing a Barlow packing and a "period 2 coloring" of one of its A_2 layers (as defined below). Alternatively, such a packing is specified by an ordered pair consisting of a doubly infinite path on a triangle 3_2 and one on a square 4_2. Just four of these packings are uniform.*

Proof. In the usual way, we find that any best packing Π_7 is determined by a best packing Π_3 in $\mathbf{R}^3$ and a 4-coloring of it. But here we have a new feature: there are infinitely many choices for Π_3.

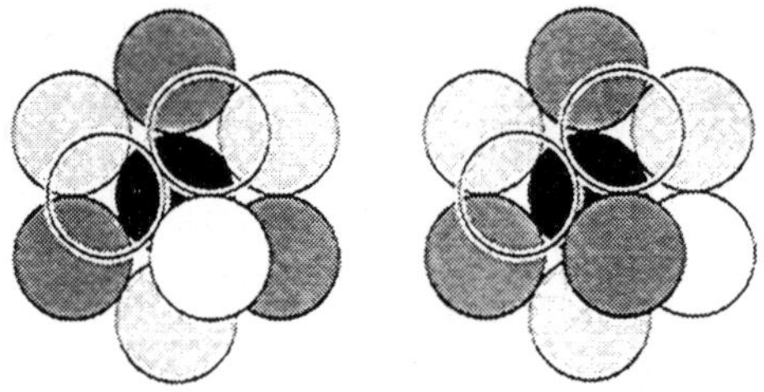

FIGURE 18. Two possible colorings of a neighborhood can be ruled out ...

However, in compensation, we find that any 4-coloring of any possible Π_3 is completely determined by its restriction to any layer (each sphere of the next layer touches three spheres in this one and so must be of the unique remaining color.)

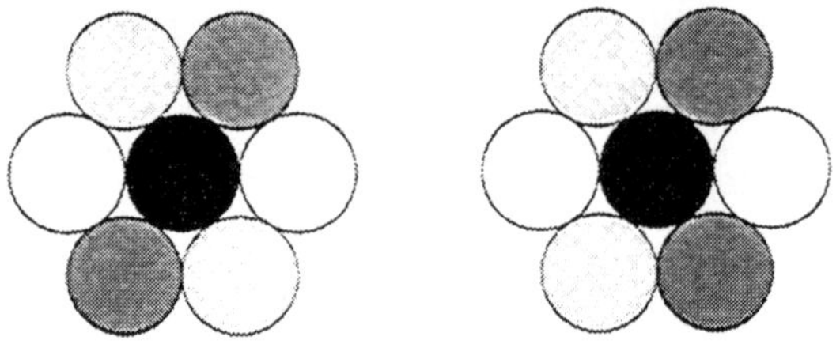

FIGURE 19. ... leaving two possibilities

Of the four ways to surround a sphere in a 4-coloring of the A_2 packing which form these layers, we find that we cannot use the cases *ababab* and *ababac*, since in each case, either choice for the next layer forces two contiguous spheres to have the same color (Figure 18). We have therefore shown that among the six neighbors of a sphere in any layer A_2 of Π_3, some pair of antipodal points has the same color, since the other two cases *abcabc* and *abacbc* have this property (Figure 19).

We next deduce that:

the A_2 can be decomposed into parallel A_1's, each of which uses just two colors.

For if every sphere in A_2 is surrounded as at left in Figure 19 then the coloring has period 2 in all directions. If any one sphere is surrounded as at right in Figure 19 then we ask: what is the pair of like-colored antipodal spheres around the white

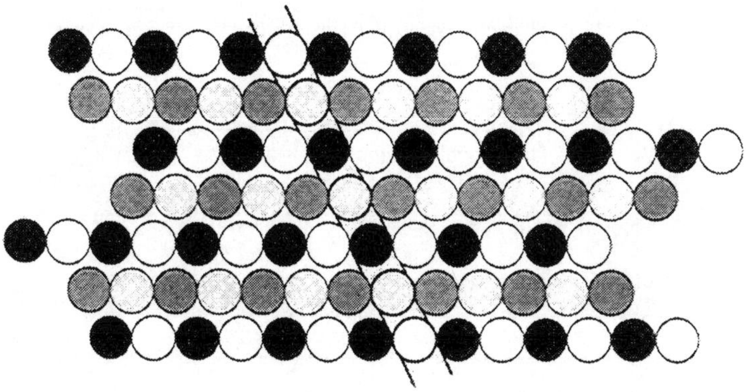

FIGURE 20. One may slide the rows; the coloring is parametrized by the colors along a strip

sphere at the right in Figure 19? The only possibility is that one of these spheres is the central black sphere. Repeating this argument, we obtain Figure 20, in which each row alternates between two of the colors.

The most general Π_7 is therefore obtained as follows: We first choose any Barlow lattice Π_3. As described before this corresponds to a choice of doubly infinite path in a triangle 3_2. We next color one row of an A_2 layer in this Π_3 with two of the four colors, say 0 and 2. We then color the next row 1 and 3, the next row after that 0 and 2, and so forth. The coloring of this layer, and consequently that of our Π_3 itself, is completely specified by a doubly infinite path on the square graph 4_2

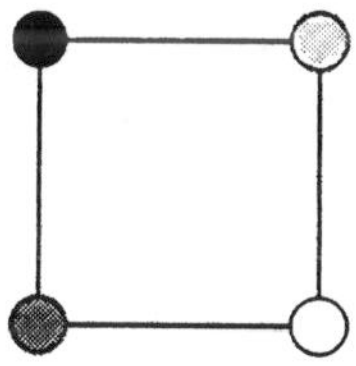

found from a sloping line such as that indicated in Figure 20. So altogether, as promised, our packing Π_7 is specified by a pair of doubly infinite paths, one in 3_2, the other in 4_2.

Moreover, since each horizontal line in our diagrams represents a translate of D_5, all such packings fiber over D_5.

There are just four cases that give uniform packings. We can extend the coloring

to either the face centered cubic lattice packing (giving Λ_7^1) or the hexagonal close packing (Λ_7^2), or the coloring

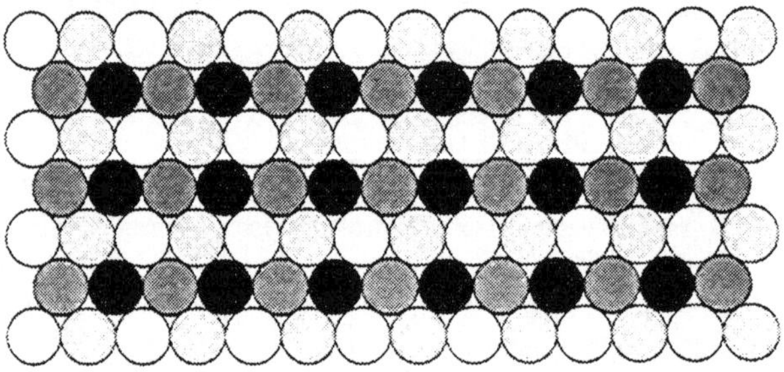

to the face centered cubic lattice packing (Λ_7^3) or the hexagonal close packing (Λ_7^4).

The first of these is the root lattice $\Lambda_7 = E_7$ and for $n = 1, 2, 3, 4$ the packing Λ_7^n consists of 1,2,2,4 translates of a lattice. □

What are all the "best" packings in $\mathbf{R}^8$?

Proposition 8 (using Hypotheses 2, 4 and 8) *The best packing in $\mathbf{R}^8$ is unique, and is the root lattice $\Lambda_8 = E_8$*

Proof. A best packing Π_8 in $\mathbf{R}^8$ is determined by a 4-coloring of a (rescaled) Π_4, equivalently, of the lattice D_4^*. We show that this coloring is unique, and in fact assigns each vector of D_4^* to the coset of D_4 it determines.

Consider the spheres of D_4^* centered at the points shown in the table. We start by arbitrarily assigning colors $0, 1, 2$ to the three contiguous spheres A, B, C. Then D must be colored 3 since it touches each of these. In a similar way, we determine the colors of all the remaining spheres in the table. It follows that this (possibly partial) coloring is invariant under permutations of the coordinates. This is because any such permutation fixes A and B, and either fixes C or takes it to another sphere like E which has been assigned the same color.

The spheres mentioned in the table, and their permutations, show that the coloring is also invariant under changing the signs of any even number of coordinates. It is similarly invariant under the subtraction of $(1, 1, 0, 0)$. Since the images of

TABLE 1. The unique 4-coloring of D_4^*

Sphere	Coordinates	Color	Touches
A	$0\ 0\ 0\ 0$	0	
B	$\frac{1}{2}\ \frac{1}{2}\ \frac{1}{2}\ \frac{1}{2}$	1	A
C	$0\ 0\ 0\ 1$	2	A, B
D	$-\frac{1}{2}\ \frac{1}{2}\ \frac{1}{2}\ \frac{1}{2}$	3	A, B, C
E	$0\ 0\ 1\ 0$	2	A, B, D
F	$-\frac{1}{2}\ -\frac{1}{2}\ \frac{1}{2}\ \frac{1}{2}$	1	A, C, D
G	$-\frac{1}{2}\ -\frac{1}{2}\ -\frac{1}{2}\ \frac{1}{2}$	3	A, C, F
H	$-1\ 0\ 0\ 0$	2	A, D, F
I	$-\frac{1}{2}\ -\frac{1}{2}\ -\frac{1}{2}\ -\frac{1}{2}$	1	A, G, H
J	$-1\ -1\ 0\ 0$	0	G, H, I

$(1,1,0,0)$ under permutations and even sign changes generate D^4, all points of any coset of D_4 are assigned the same color, and hence the coloring and the packing are unique. □

What are all the "best" packings in $\mathbf{R}^9$?

It is astonishing that although Blichfeldt completed the solution of the lattice version of the sphere packing problem in dimensions up to eight before 1930 (cf. [**CS99**]), the intervening sixty years have seen essentially no progress on the nine-dimensional problem. So it is only to be expected that the non-lattice problem will have new features in nine dimensions. We now discuss more and more surprising putatively "best" packings in nine dimensions.

Needless to say, we are nearing the end of the usefulness of our Hypotheses. But for the moment, we suppose the truth of Hypothesis 9, and use also Proposition 8, which depended in turn on Hypotheses 2, 4 and 8.

Translation

It is known [**CS99**] that the deep holes in E_8 are members of the lattice $\frac{1}{2}E_8$. It follows that if all the layers of a best packing Π_9 differ by translations, then all the corresponding placements correspond to members of $\frac{1}{2}E_8$. However, since shifting by a member of E_8 has no effect, it is better to regard the placements as members of the quotient $\frac{1}{2}E_8/E_8$.

We recall from **[CS99]** the structure of this group. There are $2^8 = 256$ cosets of E_8 in $\frac{1}{2}E_8$, and the shortest vectors in a coset are:

(i) the zero vector (1 coset),
(ii) $\pm\frac{1}{2}u$ (120 cosets),
(iii) $\pm\frac{1}{2}v_1 \pm \ldots \pm \frac{1}{2}v_8$ (135 cosets),

where $u \in E_8$ is a norm 2 vector and $v_1, \ldots, v_8 \in E_8$ are mutually orthogonal norm 4 vectors that are congruent modulo $2E_8$. The deep holes correspond to the 135 type (iii) cosets. The placement graph for this problem, therefore, has as nodes these 256 cosets, and two nodes are joined by an edge whenever the difference of the corresponding cosets is of type (iii). This does indeed have valence 135. We have thus proved:

Proposition 9A. *The best packings in* $\mathbf{R}^9$ *in which the layers are all translates of one another are parametrized by doubly infinite paths on the above graph* 256_{135}.

Rotation

If we allow adjacent layers to be related by rotations as well as translations, there are many more possibilities. To find out how many, we first consider the relationship between two adjacent layers E and F. We suppose that E is the usual E_8 defined with respect to the standard basis $e_1, \ldots, e_8$.

Each sphere of F lies above a deep hole v of E and will touch 16 spheres of E. If we take $v = e_1 = 10000000$, these will be the E spheres centered at

$$v \pm e_1, \ldots, v \pm e_8,$$

and we remark that E contains all the vectors

$$z_1e_1 + \ldots + z_8e_8$$

for which the z_i are integers with even sum, and also all vectors

$$\pm\frac{1}{2}e_1 \pm \ldots \pm \frac{1}{2}e_8$$

for which the number of minus signs is even.

However, the relation between E and F is symmetrical! There will therefore be 16 F-spheres

$$\pm f_1 \pm \ldots \pm f_8$$

touching the E-sphere centered at the origin, and F will contain all the vectors

$$v + z_1f_1 + \ldots z_8f_8$$

for which the z_i are integers with even sum, and also either all vectors

$$\pm\frac{1}{2}f_1 \pm \ldots \pm \frac{1}{2}f_8$$

for which the number of minus signs is even, or all those for which this number is odd. The vectors $\pm f_1, \ldots, \pm f_8$ must be an orthonormal basis of deep holes in E, one of which is the particular vector 10000000. Up to symmetries of E_8 it turns out that there are just four possibilities for the doubled vectors $2f_1, .., 2f_8$, as shown in Table 2.

TABLE 2. All possibilities for orthonormal bases $\pm f_1, \dots, \pm f_8$ of deep holes in E_8 containing e_1. + and − denote +1 and −1, and parentheses indicate that all cyclic permutations of the enclosed coordinates are to be applied. [The cases are named after the codes obtained by reducing their coordinates modulo 2 (see [**CP80**]).]

case	$2f_i$	group	index
d_0	(20000000)	$2^7 : S_7$	$2^0.1 = 1$
d_4	(2000) 0 0 0 0 0000(−+++)	$2^6 : S_3 \times S_4$	$2^1.35 = 70$
d_6	(20) 0 000 0 00 00(−+0++0)	$2^4 : 2^3 S_3$	$2^3.105 = 840$
e_7	(2) 0 0 00000 0(−++0+00)	$2^7 : PSL_3(2)$	$2^6.30 = 1920$

Each of these cases leads to a unique choice for F, since F may not contain the vector $\frac{1}{2}f_1 + \dots + \frac{1}{2}f_8 = \frac{1}{2}e_1 + \dots + \frac{1}{2}e_8$, which is already in E. We omit the arguments proving that these cases survive, and that the list is complete. We next compute the number of possibilities for F. For each case the table shows the structure of the group that fixes or negates the leading vector 20000000, and its index in the subgroup $2^7 : S_7$ of all automorphisms of E that do this. We conclude that the total number of such frames that contain the vector 20000000 is

$$1 + 70 + 840 + 1920 = 2831$$

and so the total number of choices for F is

$$2831 \cdot \frac{2160}{16} = 382185$$

since there are 2160 norm 4 vectors we could use in place of 20000000 and each of the frames contains 16 such vectors. From this we obtain:

Lemma 9B. *The best packings in* $\mathbf{R}^9$ *whose layers differ by arbitrary rotations and translalions are parametrized by doubly infinite paths on a placement graph of the form* ∞_{382185}.

The reason there are infinitely many possible placements is really that the automorphism group of E_8 is a maximal finite subgroup of the eight-dimensional orthogonal group, and two placements may differ by a rotation not in this subgroup. We can for instance find a best packing of $\mathbf{R}^9$ in which every pair of alternate layers is related by the matrix

$$\frac{1}{2}\begin{pmatrix} 2 & 0 & 0 & 0 & 0 & 0 & 0 & 0 \\ 0 & - & + & + & 0 & + & 0 & 0 \\ 0 & 0 & - & + & + & 0 & + & 0 \\ 0 & 0 & 0 & - & + & + & 0 & + \\ 0 & + & 0 & 0 & - & + & + & 0 \\ 0 & 0 & + & 0 & 0 & - & + & + \\ 0 & + & 0 & + & 0 & 0 & - & + \\ 0 & + & + & 0 & + & 0 & 0 & - \end{pmatrix}$$

(corresponding to case e_7), whose characteristic polynomial

$$(X-1)^2(X^2+\frac{3}{2}X+1)^3$$

shows it to be of infinite order. Since there are infinitely many placements but just 382,185 possibilities for a neighbor of a given one, the placement graph does indeed have type ∞_{382185}.

Flotation

It seems that of all these packings, only the lattice packing $\Lambda_9^1 = \Lambda_9$, was previously known. This was first described by Korkine and Zolotareff [**KZ73**] in 1873 and is their packing T_9 [**CS99**]. We describe it in some detail, because it can be modified in interesting ways.

Λ_9 consists of the vectors

$$(x_1, \ldots, x_8, 2n) \quad \text{and} \quad (x_1, \ldots, x_8+1, 2n+1),$$

where $(x_1, \ldots, x_8)$ is a typical vector of E_8 in the standard coordinate system, and n is any integer.

However, we can look at these vectors in another way! The ones with integral coordinates constitute the lattice D_9, and so

$$\Lambda_9 = D_9 \cup (D_9+v), \quad \text{where } v = \left((\frac{1}{2})^8, 0\right)$$

This leads us to ask:

Are there other vectors w for which the points of the "fluid diamond packing"

$$D_9(w) = D_9 \cup (D_9+w)$$

have minimal distance $\sqrt{2}$?

Lemma 9C. *The answer to the above question is "Yes"! In fact there are uncountably many choices for w!*

The condition is just that the squared distance of w from the nearest point of D_9 should be at least 2, and this is achieved for instance if any eight of the coordinates of w are halves of odd integers, the ninth then being completely free. Let us write

$$D_9^{\theta+} = D_9 + \left((\frac{1}{2})^8, \frac{1}{2}\theta \right).$$

Then $D_9^{0+} = \Lambda_9$ and $D_9^{1+} = D_9^{+}$ is the nine dimensional diamond packing [**CS99**].

Gold and Silver Among the Diamonds. We say that a sphere in a fluid diamond packing $D_9(w)$ is "golden" if its center is in D_9, and "silver" if its center is at a point of $D_9 + w$. Then if the squared distance of w from D_9 is strictly greater than 2, no silver sphere touches any golden one. So the packing Λ_9 has the remarkable flotation property described in the following theorem.

Flotation Theorem. *Let X and Y be two silver spheres in Λ_9. Keeping the golden spheres fixed, the silver spheres can be collectively moved so that the final position of X is the original position of Y. Usually the silver spheres do not touch the golden ones, and at no time do any two spheres ouerlap. At every stage in the motion the packing is of type $D_9^{\theta+}$ for $0 \le \theta \le 1$.*

Proof. Let $(x_1, \ldots, x_9)$ and $(y_1, \ldots, y_9)$ be the centers of X and Y.Then x_9 and y_9 are integers, but the other coordinates x_i and y_i are halves of odd integers. The desired motion is performed in 10 stages. In stage 0 we fix $x_1, \ldots, x_8$ and increase x_9 smoothly by $\frac{1}{2}$. Then at stage n, $1 \le n \le 9$, we move the nth coordinate to y_n keeping the others fixed. At all times in this motion, eight of the coordinates are halves of odd integers, and so the spheres at no time overlap. Only at the instants when some coordinate of w is an integer does any silver sphere touch a golden one. □

Unfortunately the existence of these "floating packings" violates our Hypothesis 9. We therefore simply weaken that Hypothesis to:

Hypothesis 9*. *Every best nine-dimensional packing* either *fibers over* $\Pi_8 = E_8$, or *is a fluid diamond packing.*

This is not quite so despicable as it seems, since at least Hypothesis 9* entails that every tight nine-dimensional packing is *isotopic* to one that fibers over Π_8. (Two best packings of spheres of a given radius are said to be *isotopic* if one can be continuously deformed into the other through best packings of spheres of the same radius.)

We summarize our beliefs in:

Proposition 9 (depends on Hypotheses 2, 4, 8, and 9*) *The best nine-dimensional packings are of two kinds:*

(i) Packings containing E_8, which are parametrized by random walks on the graph ∞_{382185}.

(ii) Fluid diamond packings $D_9(w)$, parametrized by the vector w.

There is an uncountably infinite collection of uniform packings.

Proof. Only the last assertion remains to be proved. The fluid diamond packing $D_9(w)$ is seen to be uniform because it has the symmetries

$$x \mapsto v + x$$

$$x \mapsto w + v - x$$

for every $v \in D_9$, and these act transitively on the spheres. □

Our real reason for believing this proposition is not that we find Hypothesis 9* inherently plausible (indeed, the flotation property initially made it seem extremely implausible), but rather that we have tried and failed many times to produce any other packings which are at least as good as those described there. The reader who finds our arguments unconvincing is invited to produce a putatively best packing not covered by the proposition!

What are the "best" packings in $\mathbf{R}^{10}$

Now we have come to the end of the useful life of the Hypotheses. It is not hard to find the best packings in dimensions $8+n$, $0 \le n \le 8$ that *do* fiber over E_8. They are parametrized by what we might call∞_{382185}-colorings of the best packings Π_n. In other words, the "color" assigned to each sphere of Π_n, is a node of the placement graph ∞_{382185} of Lemma 9B, and adjacent spheres must be colored by nodes that are adjacent in this graph. However, one such packing is the laminated lattice Λ_{10} whose density is strictly exceeded by the 10-dimensional packing found by Malcolm R. Best and briefly described below.

So we have:

Proposition 10 (depends on Hypotheses 2, 4, and 8) *Hypothesis 10 is false, even upon being weakened up to isotopy. The best known packing in* $\mathbf{R}^{10}$ *is due to Best.*

Best's packing is a uniform packing, for which Conway and Sloane have recently given a very simple construction [CS94].

The Pentacode consists of all cyclic shifts of the four vectors

$$01112, 03110, 21310, 21132,$$

and their negatives, where the digits are integers modulo 4. (These eight words are all the words of the form $c - d, b, c, d, b + c$, where b, c, d are odd.) We obtain the centers of Best's 10-dimensional packing by replacing each digit of a word of the pentacode by two integers according to the following scheme:

0 → even, even
1 → even, odd
2 → odd, odd
3 → odd, even

Our assertions about this packing follow from [CS94].

Proposition 10 suggests that an appropriate modification of Hypothesis 10 might be:

Hypothesis 10* ("Best is best"). *Best's packing is the unique best packing in 10 dimensions.*

The authors are of $2+\epsilon$ minds about the possible truth of this hypothesis.

Hypotheses 11, 12 and 13 fail as dramatically as Hypothesis 10 does. However, there may be some truth to Hypotheses 14, 15 and 16, and so some value in understanding the packings Π_{8+n} that do fiber over E_8.

Unfortunately the description in terms of ∞_{382185}-colorings of Π_n, does not make it clear *a priori* that there is more than one such Π_{8+n} However, we remark that the graph ∞_{382185}, and indeed its subgraph 256_{135} contains a copy of the complete graph 16_{15} on 16 points. This is because E_8 can be embedded in a scaled copy E_8^+ of itself having *half* the minimal norm [**CS99**]. Then E_8^+ consists of 16 cosets of E_8,and we can take those to be the desired 16 placements, since any two of them differ by a deep hole vector. This shows us that there is a particular packing Π_{8+n} corresponding to any 16-coloring of any tight packing Π_n.

We estimate that this method gives us more than 10^7 distinct, putatively best, 16-dimensional packings having the same density as the Barnes-Wall lattice Λ_{16}. This lattice Λ_{16} itself is one of these packings. It corresponds to a 16-coloring of E_8 in which the colors correspond to the cosets in E_8 of a sublattice E_8^- that is a scaled copy of E_8 at *twice* the minimal norm. The automorphisms of E_8 permute these 16 colors in just $|16.GL_4(2)| = 8.8!$ ways. So by applying all 16! color permutations we may expect to obtain at least $16!/8.8! > 10^7$ best packings Π_{16}.

Best Known Packings in Dimensions Greater Than 10

In higher dimensions, there is little reason to suspect that the best known packings are the best possible packings. We offer only one more hypothesis, Hypothesis 12* below.

For the rest of this paper we will take a quick look at some of the current best known packings in higher dimensions.

The Laminated Lattice Packings

Most of the best known packings fall into this sequence, which is defined inductively. The n-dimensional laminated lattices having a given minimum being supposed to be known, one obtains those of dimension $n+1$ by selecting the densest lattices of this dimension that contain an n-dimensional laminated lattice and have the same minimum. So for example, by laminating $\Lambda_1 = A_1$, we have the hexagonal lattice $\Lambda_2 = A_2$; laminating this we have the face-centered cubic lattice $\Lambda_3 = A_3$, and so on.

The particular cases are easiest discussed in terms of the Leech Lattice, whose discovery revolutionized this subject in 1969, so we'll define that first.

The Leech Lattice consists of the vectors in $\mathbf{R}^{24}$ whose coordinates are all congruent modulo 2, say to m, and have sum congruent to modulo 8 to $4m$, while the set of coordinates taking a given value modulo 4 is the support of a word in the binary Golay (24, 12, 8) code, for which see page 84 of [**CS99**].

We shall describe the laminated lattices of dimension up to 24 as sections of the Leech Lattice, supposing that the coordinates of the latter are grouped in a "sextet" of six tetrads the union of any two of which is a special octad—the support of a Golay codeword of weight 8. Then according to the dimension we take the vectors of the Leech Lattice that are supported on suitable subsets of the coordinates and satisfy the following conditions:

$\Lambda_1 \cong \mathbf{Z} \cong A_1$: two coordinates of the first tetrad whose sum is zero.
$\Lambda_2 \cong A_2$: three coordinates of the first tetrad whose sum is zero.
$\Lambda_3 \cong A_3 \cong D_3$: three coordinates of the first tetrad.
$\Lambda_4 \cong D_4$: the first tetrad.
$\Lambda_5 \cong D_5$: first and second tetrads, the 4 coordinates of the latter being equal.
$\Lambda_6 \cong E_6$: first and second tetrads, the last 3 coordinates of the latter being equal.
$\Lambda_7 \cong E_7$: first and second tetrads, the last 2 coordinates of the latter being equal.
$\Lambda_8 \cong E_8$: the first and second tetrads.
Λ_9: first three tetrads, the coordinates of the third being equal.
Λ_{10}: first three tetrads, the last 3 coordinates of the third being equal.
Λ_{11}: first three tetrads, the last 2 coordinates of the third being equal.
Λ_{12}: the first three tetrads.
Λ_{13}: first three tetrads, and two coordinates of the fourth having zero sum.
Λ_{14}: first three tetrads, and three coordinates of the fourth having zero sum.
Λ_{15}: first three tetrads, and three coordinates of the fourth.
Λ_{16}: the first four tetrads.
Λ_{17}: first four tetrads, and two coordinates of the fourth having zero sum.
Λ_{18}: first four tetrads, and three coordinates of the fourth having zero sum.
Λ_{19}: first four tetrads, and three coordinates of the fourth.
Λ_{20}: the first five tetrads.
Λ_{21}: all six tetrads, the 4 coordinates of the sixth being equal.
Λ_{22}: all six tetrads, the last 3 coordinates of the sixth being equal.
Λ_{23}: all six tetrads, the last 2 coordinates of the sixth being equal.
Λ_{24}: all six tetrads.

This is all summarized neatly in the Figure 21, reproduced from [**CS99**].

We should mention that in dimension 11 there are two distinct laminated lattices, and in dimension 12 and 13, three: the ones described above are those with maximal kissing number. However, the laminated lattices in all other dimensions below 25 are unique. The sequence has been continued to dimension 48, there being known to be just 23 distinct laminated lattices of dimension 25, and conjecturally many millions in dimension 26 and higher.

The alternative "Kappa" sequence is summarized in the Figure 22 reproduced from **[CS99]**. This sequence diverges from the above "Lambda" sequence in dimensions 6 through 18.

The best lattice packings known in dimensions 1 through 29 are all among those in these two sequences, as is summarized in the following chart reproduced from **[CS99]**. Note that the Kappa sequence provides denser packings only in dimensions 11, 12, and 13.

Though we cannot know the final status of the question of which packings are best in general, we feel relatively at ease proposing the following hypothesis:

Hypothesis 12* *The laminated lattice* K_{12} *is the best packing in* $\mathbf{R}^{12}$.

The History of the Laminated Lattices.

Of course the sphere packings determined by the laminated lattices in dimensions up to 3 have long been familiar, even to greengrocers. Several proofs of the optimality of the hexagonal lattice in the 2-dimensional case were offered in the nineteenth century. They are now superceded by that of Fejes Toth, which also shows that this lattice is optimal in many other senses.

The 3-dimensional laminated lattice in the face-centered cubic lattice. The question of proving its optimality is attributed to Kepler since he mentioned it in his little book "The Six-Fingered Snowflake" of 1613. In 1831 Gauss established (in a book review!) the optimality among 3-dimensional lattices, and the work of Hales on which we have already commented seems at last to have established this optimality among all 3-dimensional packings.

The St. Petersburg school of mathematicians founded by Chebyshev has had a great influence on this subject. For example, the laminated lattices in dimensions up to 9 were discovered and conjectured to be optimal by Khorkine and Zolotarev before 1873, and they and Voronoi established their optimality among lattices in dimensions up to 5. The analogous proof for dimensions 6,7 and 8 was given by Blichfeldt in the 20s and later simplified by Mordell and Vetchinkin. This is still the boundary of our positive knowledge in these matters, although in 1946 Chaundy gave a purported proof that two particular lattices were the optimal lattices in dimensions 9 and 10.

What Chaundy effectively did was to establish that the two lattices he discussed were in fact the 9- and 10-dimensional laminated lattices. In 1982, Conway and Sloane gave the formal definition of this notion of laminated lattice, and worked out all such lattices in dimensions up to 25, and the densities of all laminated lattices in dimensions up to 48. It would be incredibly difficult to continue the sequence beyond this dimension, and in any case there is little point in doing so, because denser lattices than the laminated ones are now known in the dimensions from 30 upwards.

Denser Packings than the Laminated ones.

The conjecturally "best" packings described in the first portion of this paper have the same densities as the those derived from the laminated lattices (of which they are generalizations) in all dimensions up to 9. The smallest dimension in which the laminated lattices can be beaten is 10, where they are beaten by Best's packing. This is also the smallest dimension in which a non-lattice packing appears to be optimal.

In dimensions 11,12,13 the Kappa sequence gives the best known lattices, but there are better non-lattice packings in dimensions 11 and 13, already known to the first edition of *Sphere Packings, Lattices and Groups* [**CS99**]. In all dimensions from 14 to 25 the laminated lattices (also in that book) still hold the record among lattice packings.

However, there are new, non-lattice, packings in dimensions 18 (Bierbrauer and Edel, 1998), 20 (Vardy, 1995), and 22 (Conway and Sloane, 1996). The nicest of these is the 20-dimensional packing found by Alexander Vardy in 1995. This came as a great surprise since we had considered improvements near 24 dimensions very unlikely because the Leech lattice is so spectacularly efficient there. It uses what Conway and Sloane called "the Antipode Construction", analogous to the better known "Anticode Construction" of coding theory.

We obtain Vardy's 20-dimensional packing as follows. In the Leech lattice, take all the vectors that, projected onto the 4-space spanned by the first tetrad of coordinates, fall onto one of the vertices of a regular tetrahedron of the shortest possible edge-length. Then project these vectors instead onto the orthogonally complementary 20-space. The result is not a lattice, because the sum of two vectors of the tetrahedron is not another, but it is in fact the union of four sublattices that project onto the different vertices of the tetrahedron. Conway and Sloane find its group, which is indeed obtained by projecting all the symmetries of the Leech lattice that fix the tetrahedron (as a whole) onto its orthogonal space.

They also discussed the analogous packing in 22 dimensions, where the tetrahedron in 4-space is replaced by an equilateral triangle in 2-space. It has similar properties; in particular its group is obtained from that of the Leech lattice by a rule like the one above. More recently, by a different method, Bierbrauer and Edel have found a non-lattice packing that beats the laminated lattice of dimension 18—this being the only other improvement on the first edition of *Sphere Packings, Lattices and Groups* in 25 dimensions or below. There have also been particular improvements on the upper bounds in many dimensions beyond 25, for which we refer the reader to the third edition of that book.

"Stop Press" news about lower bounds

The only new "final" result is the value of the optimal density in the 3-dimensional case. However, Henry Cohn and Noam Elkies have made some equally exciting new progress, which might soon yield the answers in 8 and 24 dimensions [**Co00**].

Some time ago, Odlyzko, Bannai and Sloane [**OdSl79**],[**BaSl81**] solved the kissing number problem in those dimensions, using an analytic method developed by Levenshtein [**Le79**]. (The only other dimensions for which that problem was solved previously were 0,1,2,3.) Cohn and Elkies' method gives a lower bound that depends on a certain function whose best value is to be determined. This function will be a linear combination of certain "radial Fourier eigenfunctions" $g_1, g_3, \dots$, and Cohn and Elkies' new bounds are achieved by finding the best combinations of the first $2n$ of these. It appears that they improve on the Rogers bound in all dimensions above 2 (where the Rogers bound is the exact answer). It also seems that in the dimensions 8 and 24 the appropriate linear combination of all the g_n will establish the optimality of the E_8 lattice and the Leech lattice, since the bounds obtained by taking larger and larger finite numbers of the g_n appear to be tending to the correct limits.

If so, all that will be required to establish these two answers will be the "divination" of the function corresponding to the correct infinite linear combination of the g_n, and the verification that it satisfies certain inequalities. So we might soon have a solution in 8 and 24 dimensions that is very much shorter than the one in 3 dimensions! But since this depends on somebody's having a bright idea it is hard to know just how long it will take.

All our other so-called "progress" is at the level of conjecture. But we can now say that we *think* we know all the "best" packings in dimensions 1-10, and although this is merely a conjectural answer to an undefined question, this is a kind of "knowledge" that we didn't previously have!

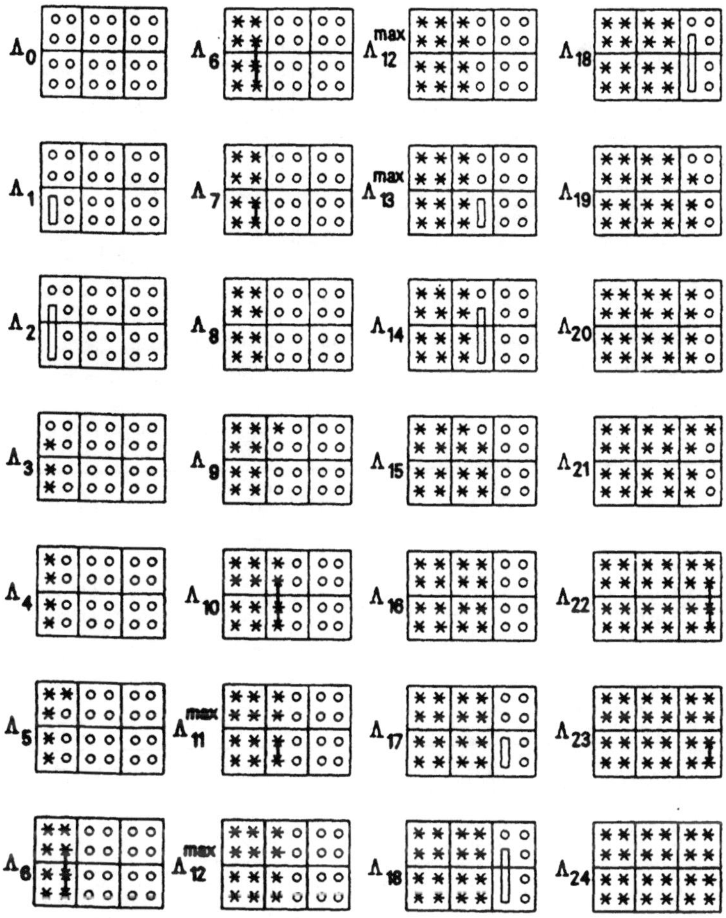

FIGURE 21. Coordinates for $\Lambda_0, \ldots, \Lambda_n$ as sections of the Leech Lattice. A small circle represents a zero coordinate, a hollow loop is a set of coordinates adding to zero, and asterisk is a free coordinate, and a line of asterisks is a set of equal coordinates.

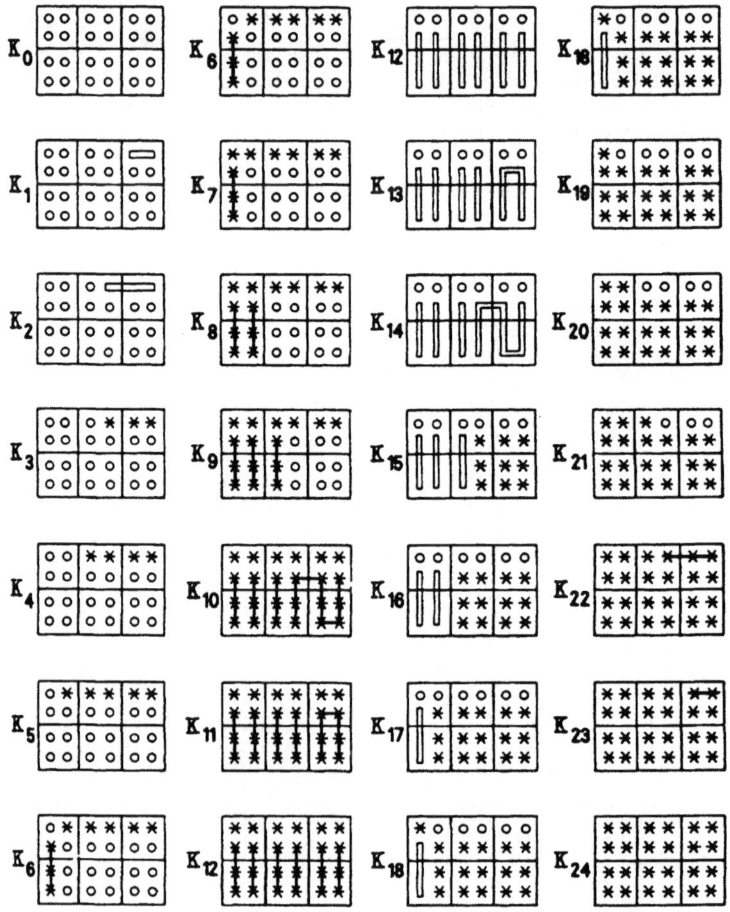

FIGURE 22. Coordinates of the lattices $K_0, \ldots, K_n$ as sections of the Leech Lattice.

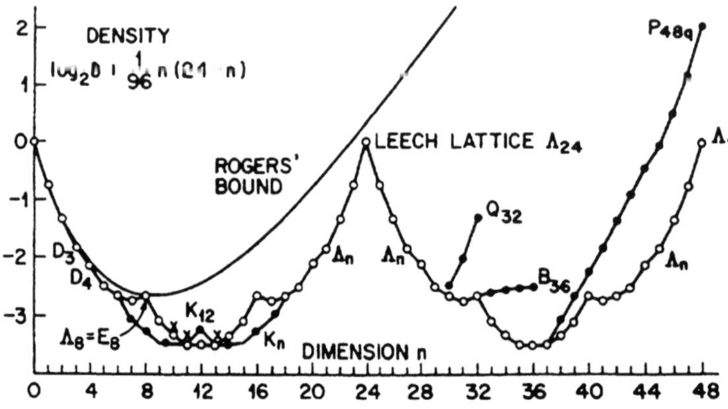

FIGURE 23. The densest sphere packings known in dimensions $n \leq 48$. See [CS99].

References

[BaSl81] E. Bannai and N. J. A. Sloane, Uniqueness of certain spherical codes, *Can. J. Math.*, **33**, (1981), 437-449 ≅ Chap. 14 of [**CS99**].

[Ba83] W. Barlow, Probable nature of the internal symmetry of crystals, *Nature*, **29** (1883), 186–188.

[Ba59] E. S. Barnes, The construction of perfect and extreme forms II, *Acta Arith.*, **5** (1959), 205–222.

[Be80] M. R. Best, Binary codes with a minimum distance of four, *IEEE Trans. Inform. Theory* **26** (1980), 738–742.

[Ch46] T. W. Chaundy, The arithmetic minima of positive quadratic forms, *Quart. J. Math.*, **17** (1946), 166–192.

[Co00] H. Cohn, Ph.D. dissertation, Harvard (2000).

[Co69] J. H. Conway, Personal communication to J. Leech, 1969.

[CP80] J. H. Conway and V. Pless, On the enumeration of self-dual codes, *J. Combin. Theory*, Series A, **28** (1980), 26–53. (See also the corrections in J. H. Conway, V. Pless and N. J. A. Sloane, The binary self-dual codes of length up to 32: a revised enumeration, *J. Combin. Theory*, Series A, **60** (1992), 183–195.)

[CS94] J. H. Conway and N. J. A. Sloane, Quaternary constructions for the binary single-error-correcting codes of Julin, Best and others, *Designs, Codes and Cryptography*, **41** (1994), 31–42.

[CS95] J. H. Conway and N. J. A. Sloane, What are all the best sphere packings in low dimensions?, *Disc. Comb. Geom.*, **13** (1995), 383-403.

[CS96] J. H. Conway and N. J. A. Sloane, The antipode construction for sphere packings, *Invent. Math.*, **123** (1996), 309-313.

[CS99] J. H. Conway and N. J. A. Sloane, *Sphere Packings, Lattices and Groups*, Third edition, Springer-Verlag, New York, 1999.

[CT53] H. S. M. Coxeter and J. A. Todd, An extreme duodenary form, *Canad. J. Math.*, **5** (1953), 384–392.

[CR51] H. M. Cundy and A. P. Rollett, *Mathematical Models*, Oxford Univ. Press, 1951.

[GFT83] G. Fejes Tóth, New results in the theory of packing and covering, pp. 318–359 of P. M. Gruber and J. M. Wills, editors, *Convexity and Its Applications*, Birkhäuser, Basel, 1983.

[LFT68] L. Fejes Tóth, Solid circle-packings and circle-coverings, *Studia Sci. Math. Hungar.*, **3** (1968), 401–409.

[Ha92] T. C. Hales, The sphere packing problem, *J. Computational and Applied Math.*, **44** (1992), 41–76.

[Ha94] T. C. Hales, The status of the Kepler conjecture, *Mathematical Intelligencer*, **16**(3) (1994), 47-48.

[Ha97a] T. C. Hales, Sphere Packings I, *Disc. Comp. Geom.*, **17** (1997), 1-51.

[Ha97b] T. C. Hales, Sphere Packings II, *Disc. Comp. Geom.*, **18** (1997), 135-149.

[Ha97c] T. C. Hales, Sphere Packings III, preprint.

[Ha98] T. C. Hales, The Kepler Conjecture, *preprint.*

[Hs93] W.-Y. Hsiang, On the sphere packing problem and the proof of Kepler's conjecture, *Int. J. Math.*, **4** (1993), 739–831.

[KZ73] A. Korkine and G. Zolotareff, Sur les formes quadratiques, *Math. Ann.*, **6** (1873), 366–389.

[Le67] J. Leech, Five-dimensional nonlattice sphere packings, *Canad. Math. Bull.*, **10** (1967), 387–393.

[Le69] J. Leech, Six and seven dimensional nonlattice sphere packings, *Canad. Math. Bull.*, **12** (1969), 151–155.

[LS70] J. Leech and N. J. A. Sloane, New sphere packings in dimensions 9–15, *Bull. Amer. Math. Soc.*, **76** (1970), 1006–1010.

[LS71] J. Leech and N. J. A. Sloane, Sphere packing and error-correcting codes, *Canad. J. Math.*, **23** (1971), 718–745. A revised version appears as Chap. 5 of [**CS99**].

[Le79] V. I. Levenshtein, On bounds for packings in n-dimensional Euclidean space, *Doklady Akad. Nauk SSR*, **245** (1979), 1299-1303 = *Sov. Math. Doklady*, **20** (1979), 417-421.

[Me47] S. Melmore, Densest packing of equal spheres, *Nature*, **159** (1947), 817.

[Mu93] D. Muder, A new bound on the local density of sphere packings, *Discrete Comp. Geom.*, **10** (1993), 351–375.

[Ne98] G. Nebe, Some cyclo-quaternionic lattices, *J. Alg.* , **199** (1998), 472-498.

[OdSl79] A. M. Odlyzko and N. J. A. Sloane, New bounds on the number of unit spheres that can touch a unit sphere in n dimensions, *J. Comb. Th., Ser. A*, **26**, 210-214 ≅ Chap. 13 of [**CS99**].

[Ro64] C. A. Rogers, *Packing and Covering*, Cambridge Univ. Press, 1964.

[Va99] A. Vardy, Density doubling, double circulants, and new sphere packings, *Trans. A. M. S.*, **351** (1999), 271-283.

MATHEMATICS DEPARTMENT, PRINCETON UNIVERSITY, PRINCETON, NEW JERSEY 08544,

INSITITUTO DE MATEMATICAS, U.N.A.M., A.P. 273 ADMON. DE CORREOS # 3, CUERNAVACA, C.P. 62251, MORELOS, MEXICO

MATHEMATICAL SCIENCES RESEARCH CENTER, AT&T BELL LABORATORIES, MURRAY HILL, NEW JERSEY 07974,

A report on the proof of the Langlands conjectures for $GL(N)$ over p-adic fields

Guy Henniart

In November 1999, I was invited to Cambridge, Mass. to give a talk at the conference on "Current developments in Mathematics". I lectured then on the work of Michael Harris and Richard Taylor [Ha T] on Shimura varieties, which provided in particular a proof of the Langlands conjectures for GL(N) over p-adic fields. I had written a preconference report, essentially a written account of the intended talk : you will find it below. Then it was my duty to write a more extensive report for the definitive proceedings. However there is already a very nice review by H. Carayol at the Bourbaki seminar [Ca]. It gives a rather extensive exposition of the geometric part of the work of Harris and Taylor, and also applications. Not being a specialist of algebraic geometry or Shimura varieties, I did not feel like duplicating, let alone emulating, Carayol. But it seemed to me that a report on the local-global principles behind the proof of the Langlands conjectures could be helpful, and I could talk there of things which are not very much explained elsewhere. Needless to say the reader should also consult [Ca] and [Ha T], especially the introduction, to get a more balanced overview. Laumon's talk at the same conference, on Lafforgue's work on the global Langlands conjectures for GL_n over function fields, is also relevant here.

I have kept the notes written for the talk as a first part A, thinking it provided a gentle general introduction to the topics to follow. On the other hand, I have taken the liberty of discussing in part B some more recent results like the proof of the Shimura-Taniyama-Weil conjecture, or the proof of the Artin conjecture for infinitely many degree 2 iscosahedral representations of the absolute Galois group of $\mathbb{Q}$.

Let me finally take the opportunity to thank the organizers of the "Current Developments in Mathematics" series for the invitation, M. Harris and R. Taylor for communications on the subject, and, above all, R.P. Langlands for inspiration.

Main references :

- [Ca] **H. CARAYOL.** — *Preuve de la conjecture de Langlands locale pour* GL_n *: travaux de Harris-Taylor et Henniart*, Sém. Bourbaki *n°* 857, March 1999, Astérisque *n°* 266, SMF, Paris (2000), 191-244.
- [Ha 1] **M. HARRIS.** — *Supercuspidal representations in the cohomology of Drinfeld's upper half-spaces ; elaboration of Carayol's program*, Invent. Math. 129 (1997), 75-120.
- [Ha 2] **M. HARRIS.** — *The local Langlands conjecture for* GL_n *over a p-adic field*, $n < p$, Invent. Math. 134 (1998), 177-210.
- [Ha T] **M. HARRIS, R.TAYLOR.** — *On the geometry and cohomology of some simple Shimura varieties*, Preprint *n°* 227, Institut de Mathématiques de Jussieu, Paris, octobre 1999.
- [He] **G. HENNIART.** — *Une preuve simple des conjectures de Langlands pour* GL_n *sur un corps p-adique*, Invent. Math. 139 (1999), 439-455.

A) A general overview

As mentioned in the introduction, this is the account which I wrote for the conference, with only cosmetic changes. See also Carayol's report [Ca], my report at the Journées Arithmétiques [He JA] – and of course [Ha T].

A I) The main results, and the local-global principle

1. Introduction

The conjectures of Langlands that I will talk about relate two *a priori* rather different kinds of objects : Galois representations and automorphic representations. The conjectures come in two versions : global conjectures concern number fields, that is finite extensions of $\mathbb{Q}$, whereas local conjectures concern their completions, p-adic fields. In July 1998, M. Harris and R. Taylor announced a proof of the local conjectures, and I shall report on their proof. The principle is a local-global one : a weak result is obtained over some number fields, but this gives enough information for the full local result. The method is geometric : the two kinds of objects are found in the cohomology of some specially constructed algebraic varieties ; as a byproduct, one gets a geometric model for the local correspondence.

There are also versions of the conjectures over fields of positive characteristics. There the local result has been known for some time, and very spectacular results on the global correspondence were recently obtained by L. Lafforgue. I refer you to Laumon's talk at this conference.

2. Galois representations

Let F be a number field, $\overline{F}$ an algebraic closure, G_F the Galois group of $\overline{F}$ over F. Galois representations are (continuous) homomorphisms $G_F \to \mathrm{Aut}(V)$ where V is some finite-dimensional vector space. There are two main sources of Galois representations.

The first comes upon taking a finite Galois extension E of F in $\overline{F}$ and representing the Galois group $\mathrm{Gal}(E/F)$ in some finite dimensional complex vector space. As $\mathrm{Gal}(E/F)$ is a quotient of G_F, one gets a **complex** representation of G_F. Since G_F is the inverse limit of such finite Galois groups $\mathrm{Gal}(E/F)$, studying those complex representations amounts to studying G_F, so to say, on the Fourier transform side. It is known often to be a good idea.

The second supply of Galois representation is geometry. Let X be a smooth projective variety over F. Then G_F acts on $X \otimes \mathrm{Spec}(\overline{F})$, and on the ℓ-adic etale cohomology groups $H^i(X \otimes \mathrm{Spec}(\overline{F}), \mathbb{Q}_\ell)$ for any prime number ℓ. Those are finite dimensional $\mathbb{Q}_\ell$-vector spaces. It is good to allow extension of coefficients to an algebraic closure $\overline{\mathbb{Q}}_\ell$ of $\mathbb{Q}_\ell$, for example to decompose those cohomology spaces under the action of G_F, so one gets **ℓ-adic** representations $G_F \to \mathrm{Aut}(V)$, where V is a finite dimensional vector space over $\overline{\mathbb{Q}}_\ell$. Contrary to the complex representations above, their image is usually infinite. For example, one can take an elliptic curve over F and look at the representation of G_F on its Tate module at a prime number ℓ; this gives a 2-dimensional ℓ-adic representation of G_F.

Both types of representations – complex or ℓ-adic – share a crucial common property : they are **unramified almost everywhere**. Let me explain this; one looks at the different non-archimedean completions ("finite places") F_v of F. Choosing an extension to $\overline{F}$ of the completion yields an algebraic closure $\overline{F}_v$ of F_v and an embedding of the Galois group G_{F_v} of $\overline{F}_v$ over F_v into G_F; choosing a different extension gives a conjugate embedding. Each completion has a residue field and the residue field of F_v is a finite field, having the residue field of $\overline{F}_v$ as an algebraic closure. The group G_{F_v} has an inertia subgroup I_v, made out of the elements acting trivially on the residue field. The quotient G_{F_v}/I_v is a free profinite group with one prefered generator Frob_v, acting as $x \mapsto x^{1/q_v}$ on the residue field of $\overline{F}_v$, where q_v is the cardinality of the residue field of F_v.

Let $\Sigma : G_F \to \mathrm{Aut}(V)$ be a complex or ℓ-adic representation as above. Then for all but a finite number (we say "almost all") of finite places v of F, Σ is **unramified**, that is, trivial on I_v. The semisimplification of the restriction Σ_v of Σ to G_{F_v} is then completely characterized by the local L factor

$$L(\Sigma_v, s) = \det(1 - \Sigma(\mathrm{Frob}_v)q_v^{-s})^{-1}$$

seen, say, as a Dirichlet series.

Moreover, if Σ itself is semisimple, then it is determined by the data of $L(\Sigma_v, s)$ at almost all v's.

3. Class-field theory

Before stating the Langlands conjectures for GL_n, I have to recall the case $n = 1$, given by class field theory. One-dimensional complex representations of G_F, up to isomorphism, are simply characters (i.e. continuous homomorphisms) $G_F \to \mathbb{C}^\times$ (they have finite order). Class field theory gives a way to describe them all, purely in terms of the base field F. One needs the ring of **adeles** $\mathbb{A}_F$ of F, the subring of $\prod F_v$ (the product is over all places v of F, including archimedean completions) made out of families (x_v) where x_v is in the ring of integers O_{F_v} for almost all v's. The group of units $\mathbb{A}_F^\times = GL_1(\mathbb{A}_F)$ carries a natural locally compact topology, and $F^\times$ sits diagonally in it as a discrete subgroup.

A character χ of $\mathbb{A}_F^\times$ is a continuous homomorphism into $\mathbb{C}^\times$. By continuity it is unramified (i.e. its restriction to $F_v^\times \subset \mathbb{A}_F^\times$ is trivial on the unit group of $F_v^\times$) for almost all finite v's; at such places its restriction to $F_v^\times$ is completely characterized by its value on a uniformizer ϖ_v of F_v.

One way to state class field theory is to say that it gives a one-to-one correspondence $\chi \mapsto \pi(\chi)$ between characters of G_F and finite order characters of $\mathbb{A}_F^\times$ **trivial on** $F^\times$. That correspondence is characterized by the fact that at almost all v's, where χ and $\pi(\chi)$ are unramified, one has

$$\pi(\chi)(\varpi_v) = \chi(\mathrm{Frob}_v).$$

An important feature is that the correspondence $\chi \leftrightarrow \pi(\chi)$ is determined by looking only at "good" unramified places. What happens at the other "bad" places is then forced, so one can hope to obtain something back at those bad places. This is indeed the case and gives local class field theory. If v is a finite place of F, there is a one-to-one correspondence $\chi \mapsto \pi(\chi)$ between characters of G_{F_v} and finite order characters of $F_v^\times$. That correspondence is the only one compatible with the global one in the sense that if χ is a character of G_F, then for any finite place v of F, $\pi(\chi_v)$ is the restriction of $\pi(\chi)$ to $F_v^\times$. But the local correspondence depends only upon F_v and not on the way we see it as a completion. This is what I call the **local-global principle of class-field theory**. We shall see it can be generalized to the context of the Langlands conjectures.

Note : If W_{F_v}, the Weil group of F_v, is the subgroup of G_{F_v} made out of elements acting on the residue field as an integral power of Frob_v, then there is a one-to-one correspondence between characters of W_{F_v} and (all) characters of $F_v^\times$. In the p-adic case it is often more convenient to work with the Weil group.

4. Automorphic representations

To generalize the above result from one-dimensional to n-dimensional Galois representations (where n is any integer ≥ 2), we need something on the other side. This is provided by **automorphic** forms and representations, which are an adelic generalization of classical modular forms and the action on them of Hecke operators.

If F is a number field as above, we can form the group $\mathrm{GL}_n(\mathbb{A}_F)$; as for $n = 1$ it carries a locally compact topology and $\mathrm{GL}_n(F)$ sits inside diagonally as a discrete subgroup. The group $\mathrm{GL}_n(\mathbb{A}_F)$ acts on $\mathrm{GL}_n(F)\backslash\mathrm{GL}_n(\mathbb{A}_F)$ by right translations. Automorphic forms are complex functions on $\mathrm{GL}_n(F)\backslash\mathrm{GL}_n(\mathbb{A}_F)$ with various smoothness and finiteness assumptions; the group $\mathrm{GL}_n(\mathbb{A}_F)$ acts on automorphic forms. To define the representations we are interested in, it is convenient to fix a character (unitary, say) ω of $F^\times\backslash\mathbb{A}_F^\times$ and look only at those functions where the centre of $\mathrm{GL}_n(\mathbb{A}_F^\times)$ (isomorphic to $\mathbb{A}_F^\times$) acts via multiplication by ω. The space $L^0(\omega)$ of cuspidal automorphic forms is then the space of functions on $\mathrm{GL}_n(F)\backslash\mathrm{GL}_n(\mathbb{A}_F)$, transforming by ω under the centre, square integrable modulo the centre, and satisfying the cuspidality condition : for any proper parabolic subgroup P of GL_n over F with unipotent radical N, one has

$$\int f(ng)dn = 0 \text{ almost everywhere},$$

where the integral is over $N(F)\backslash N(\mathbb{A}_F)$. (This is the generalization of the classical notion of parabolic modular form). The space $L^0(\omega)$ is the direct sum of irreducible inequivalent unitary representations of $\mathrm{GL}_n(\mathbb{A}_F)$. Such summands (for varying ω) are called **automorphic cuspidal representations** of $\mathrm{GL}_n(\mathbb{A}_F)$.

Let Π be such an automorphic cuspidal representation of $\mathrm{GL}_n(\mathbb{A}_F)$. Then due to the topology of $\mathrm{GL}_n(\mathbb{A}_F)$, Π appears as a kind of completed tensor product $\otimes\Pi_v$, where for each finite place v, Π_v is a **smooth irreducible** representation of $\mathrm{GL}_n(F_v)$ i.e. a representation in some complex vector space, usually infinite dimensional, which is irreducible in the algebraic sense and verifies the strong continuity condition that each vector has an open stabilizer in $\mathrm{GL}_n(F_v)$. (At infinite places, the notion is slightly different but I shall speak abusively of smooth irreducible representations also at those places).

For almost all finite places v, Π_v is **unramified**, i.e. has a non-zero vector fixed under $\mathrm{GL}_n(\mathcal{O}_{F_v})$. Such an unramified Π_v is characterized by its L-factor $L(\Pi_v, s)$ which is of the form $P(q_v^{-s})^{-1}$, P a degree n complex polynomial such that $P(0) = 1$.

The automorphic cuspidal representation Π itself is determined by the data $L(\Pi_v, s)$ at almost all (unramified) places.

5. The global Langlands conjectures

Let $n \geq 2$ be an integer. We say that an n-dimensional **irreducible** complex Galois representation Σ of G_F and an automorphic cuspidal representation Π of $\mathrm{GL}_n(\mathbb{A}_F)$ **correspond** if for almost all finite v's, we have $L(\Pi_v, s) = L(\Sigma_v, s)$. The hope expressed by the Langlands conjectures is that there are many corresponding pairs (Σ, Π).

If Σ is an irreducible **complex** representation of G_F, then certainly there should be a corresponding Π, but the Π's thus obtained are very special. An automorphic cuspidal representation Π of $\mathrm{GL}_n(\mathbb{A}_F)$ will not generally correspond to a complex Galois representation. However, if the components of Π at infinity satisfy some algebraicity condition (smooth irreducible representations at the infinite places have been classified by Langlands), then Π should have, for any prime number ℓ, a corresponding ℓ-adic Galois representation of G_F. More precisely there should exist a half integer k (depending on n and the components of Π at infinite places), a number field E in $\mathbb{C}$ such that for almost all finite v, $L(\Pi_v, s+k)$ belongs to $E(q_v^{-s})$ and, for each finite place λ of E, a continuous homomorphism $\Sigma_\lambda : G_F \to \mathrm{GL}_n(E_\lambda)$ (where E_λ is the completion of E at λ), which, at almost all places v, is unramified and verifies

$$L(\Pi_v, s+k) = L(\Sigma_{\lambda,v}, s).$$

In the other direction, it is generally accepted that every ℓ-adic irreducible Galois representation coming from geometry as in section 2 should correspond to an automorphic cuspidal representation, in the sense just explained. An instance is provided by the recently proven theorem that every elliptic curve over $\mathbb{Q}$ is modular : indeed the statement is equivalent to saying that the ℓ-adic representation of $G_{\mathbb{Q}}$ on the Tate module of the elliptic curve corresponds to some automorphic cuspidal representation of GL_2 over $\mathbb{Q}$.

6. L-functions

Class field theory describes all finite abelian extensions of a number field F purely in terms of the base field F, using adeles, or, what amounts to the same, ideals and congruences. It is more difficult to say what is gained when a given Galois representation (of dimension $n \geq 2$) is shown to correspond to an automorphic cuspidal representation. One tangible gain is information on L-functions, to which we turn now.

If Π is an automorphic cuspidal representation of GL_n over F, then its L-factors are in fact defined at every place, and give rise to a global L-function with nice analytic properties. More precisely, let F_0 be a p-adic field, and $\psi_0 : F_0 \to \mathbb{C}^\times$ a non-trivial additive character. Any smooth irreducible representation π of $\mathrm{GL}_n(F_0)$ has an L-factor $L(\pi, s)$ of the form $P(q_0^{-s})^{-1}$ where q_0 is the cardinality of the residue field of F_0 and P some complex polynomial of degree $\leq n$ with value 1 at 0; one also defines a factor $\varepsilon(\pi, s, \psi_0)$ which is a non-zero monomial in q_0^{-s}. There are also such definitions for archimedean fields such as $\mathbb{R}$ and $\mathbb{C}$. If Π is global as above, then the product $L(\Pi, s) = \prod_v L(\Pi_v, s)$ (over all places v) converges for $\mathrm{Re}(s)$ big enough and extends to a meromorphic function of s (it is even an **entire** function because we have assumed $n \geq 2$) which satisfies a functional equation analogous to the functional equation for Riemann's zêta function

$$L(\Pi, s) = \varepsilon(\Pi, s)L(\Pi^\vee, 1-s).$$

(Here, and in the following, a superscript $^\vee$ indicates the dual or "contragredient" representation; indeed there is a natural notion of dual representation in those contexts). If we fix a non-trivial character Ψ of $\mathbb{A}_F$ trivial on F, the factor $\varepsilon(\Pi, s)$ factors as $\coprod \varepsilon(\pi_v, s, \Psi_v)$, where almost all factors are identically 1 and Ψ_v is the component of Ψ at $v : \Psi_v : F_v \to \mathbb{A}_F \to \mathbb{C}^\times$.

So if an irreducible Galois representation Σ of G_F is shown to correspond to an automorphic cuspidal representation Π of GL_n over F, then one can complete the definition of L-factors of Σ from unramified places to all places, and get a global L-function for Σ with very nice analytic properties.

It is not straighforward to guess the bad factors $L(\Sigma_v, s)$ just by looking at Σ_v; indeed, if Σ is an ℓ-adic representation, the L-factors at infinite places presumably involve looking at the restriction of Σ to G_{F_λ}, where λ runs through **ℓ-adic** places of F. However, if Σ is complex and v is any place of F, or if Σ is ℓ-adic and v is a finite place of F prime to ℓ, then there is an intrinsic definition of $L(\Sigma_v, s)$ (determined by the local representation Σ_v) which goes back to E. Artin, and Langlands and Deligne have shown how to define local factors $\varepsilon(\Sigma_v, s, \psi_v)$ for any non-trivial additive character ψ_v of F_v.

It is expected, for example, that if Σ is an irreducible complex representation of G_F of degree ≥ 2 corresponding to an automorphic cuspidal representation Π, then the L and ε-factors of Σ and Π coincide at all places. In any case, the Artin L-function $L(\Sigma, s) = \prod_v L(\Sigma_v, s)$ is then entire, yielding **Artin's conjecture** for Σ.

Remark. There is even a notion of L-factors for pairs, which will be important in the next section. If F_0 is a p-adic field and $\psi_0 : F_0 \to \mathbb{C}^\times$ a non-trivial character, and if π, π' are smooth irreducible representations of $\mathrm{GL}_n(F_0)$ and $\mathrm{GL}_{n'}(F_0)$ respectively, then one defines factors $L(\pi \times \pi', s)$ and $\varepsilon(\pi \times \pi', s, \psi_0)$. If π, π' are unramified and $\quad L(\pi, s) = \prod_i (1 - \alpha_i q_0^{-s})^{-1}, \qquad L(\pi', s) = \prod_j (1 - \beta_j q_0^{-s})^{-1}$ then $L(\pi \times \pi', s) = \prod_{i,j}(1 - \alpha_i\beta_j q_0^{-s})^{-1}$. Similar definitions exist for archimedean fields. If π' is the trivial character of $\mathrm{GL}_1(F_0)$ one recovers the previous definition of factors for π. If Π, Π' are automorphic cuspidal representations of GL_n and $\mathrm{GL}_{n'}$ respectively over a number field F, then the product over all places gives a meromorphic function $L(\Pi \times \Pi', s)$ (there are poles, but they are known) and a monomial $\varepsilon(\Pi \times \Pi', s) = \alpha N^s$ for some $\alpha \in \mathbb{C}^\times$ and $N \in \mathbb{N}$. There is a functional equation

$$L(\Pi \times \Pi', s) = \varepsilon(\Pi \times \Pi', s)L(\Pi^\vee \times \Pi'^\vee, 1-s)$$

If Π and Π' happen to correspond to complex Galois representations Σ and Σ' one expects of course the equalities

$$L(\Pi \times \Pi', s) = L(\Sigma \otimes \Sigma', s)$$
$$\varepsilon(\Pi \times \Pi', s) = \varepsilon(\Sigma \otimes \Sigma', s)$$

as it is the case for the components at unramified places : they should be the consequence of the analogous equalities for the components at the other places.

7. From global to local : the results

A local-global principle is emerging, generalizing the principle already present in class field theory. The basic idea is the following. The geometry of some special varieties introduced by Shimura yields correspondings pairs (Σ, Π) over some number fields F (in this geometric context, Σ is rather an ℓ-adic representation). If v is a finite place of F, prime to ℓ, one expects that the restriction Σ_v of Σ to G_{F_v} depends only on the component Π_v of Π at v, and conversely. Besides the local L and ε factors of Π_v and Σ_v should correspond. Moreover one hopes that all local representations occur that way. This is essentially what Harris and Taylor prove.

The first thing is to get corresponding pairs, matching at almost all places. The following is an improvement, due to Harris and Taylor, of a theorem of Clozel. To state it conveniently it is better to fix a prime number ℓ and an isomorphism of $\overline{\mathbb{Q}}_\ell$ with $\mathbb{C}$.

THEOREM 1 (Global correspondence I). — *Let F be a* CM*-field with complex conjugation c. Let n be a positive integer and Π an automorphic cuspidal representation of* GL_n *over F. Assume*

(1) *The transform Π^c of Π by complex conjugation is isomorphic to $\Pi^\vee$*

(2) *Π is algebraic and regular enough at infinite places*

(3) *at some finite places, the matrix coefficients of Π_v are L^2-modulo the centre.*

Then there is a positive integer a and an ℓ-adic representation $\Sigma' : G_F \to \mathrm{GL}_{an}(\overline{\mathbb{Q}}_\ell)$ such that at almost all finite places v of F

$$L(\Sigma'_v, s) = L(\Pi_v, s + \frac{n-1}{2})^a .$$

Note. Condition (2) says more precisely (and technically) that Π_∞, the tensor product of the Π_v for v infinite, has the same infinitesimal character as some algebraic finite dimensional complex representation of $\mathrm{Res}_{F/\mathbb{Q}}(\mathrm{GL}_n)$. Condition (3) is a technical condition, imposed by our present state of knowledge. A matrix coefficient for Π_v is a function on $\mathrm{GL}_n(F_v)$ of the form $g \mapsto \lambda(\Pi_v(g)x)$ where x is a vector in the space of Π_v and λ a linear functional on that space which is smooth with respect to the natural action of $\mathrm{GL}_n(F_v)$.

Of course, one expects that $\Sigma' = a\Sigma$ for some ℓ-adic representation $\Sigma : G_F \to \mathrm{GL}_n(\overline{\mathbb{Q}}_\ell)$.

In fact Harris and Taylor first prove some special cases of the theorem above, which are enough for the expected local results; the theorem in its generality comes as an afterthought. Moreover, it can be **reinforced** by a statement at all finite places v of F prime to ℓ. Take such a place, possibly bad for Π i.e. with Π_v not unramified. Looking at the bad reduction of Shimura varieties at such a place (see Lecture II), Harris and Taylor show that if not Σ'_v then at least its semisimplification is of the form $a\Sigma_v$ where Σ_v is determined by Π_v and can be read from Π_v via a geometric local model. Seeing a given p-adic field F_0 as a completion of some suitable CM field, one gets a purely local result.

THEOREM 2 (Local correspondence). — *Let F_0 be a p-adic field and $\psi_0 : F_0 \to \mathbb{C}^\times$ a non-trivial additive character. Then for each positive integer n, there is a canonical map $\sigma \mapsto \pi(\sigma)$ between irreducible degree n complex representations* *of* W_{F_0}*, up to isomorphism, and smooth irreducible representations of* $\mathrm{GL}_n(F_0)$*, up to isomorphism. Those maps are injective and if* σ*,* σ' *are two irreducible complex representations of* W_{F_0}*, of degree* n*,* n' *respectively, then one has*

$$L(\pi(\sigma) \times \pi(\sigma'), s) = L(\sigma \otimes \sigma', s)$$
$$\varepsilon(\pi(\sigma) \times \pi(\sigma'), s, \psi_0) = \varepsilon(\sigma \otimes \sigma', s, \psi_0).$$

In fact the image is known : it consists of the supercuspidal representations (their matrix coefficients have compact support modulo the centre). Also the maps $\sigma \mapsto \pi(\sigma)$ can be characterized by the preservation of L and ε-factors above, and some other milder natural requirements, like $\pi(\sigma^\vee) = \pi(\sigma)^\vee$.

The inverse map of $\sigma \mapsto \pi(\sigma)$, written $\pi \mapsto \sigma(\pi)$, can be extended to a map $\pi \mapsto \sigma(\pi)$ which to each smooth irreducible representation of $\mathrm{GL}_n(F_0)$ associates a semisimple representation of W_{F_0} (this is no longer injective; to restore bijectivity one has to complicate things further by considering the Weil-Deligne group ...).

Thm. 2 proves the Langlands conjectures for GL_n over F_0, but Harris and Taylor do more : they provide a geometric local model for the correspondence. But I shall try and explain that Part A II. As mentioned above, Thm. 2, in turn, has global consequences.

THEOREM 3 (Global correspondence II). — *In the situation of theorem 1 above, let* v *be any finite place of* F *prime to* ℓ*. Then the semisimplification of the restriction of* Σ_v *to* W_{F_v} *is isomorphic to* $\sigma(\Pi_v)$*.*

There is some hope to determine Σ_v fully (up to isomorphism) from Π_v.

Remark. The methods used by Harris and Taylor are inspired by Deligne, who started the case $n = 2$. The theorems above when $n = 2$ were obtained by Carayol.

A II) The role of the geometry of Shimura varieties

8. Introduction

In this second part, I shall try to describe roughly how the geometry of Shimura varieties is used to get the main results. Of necessity, this part is more technical (even) that the previous one. I will not be able to give much detail, and one should not take the statements too literally. I simply hope to convey a roughly resembling picture of the methods and steps involved. Again, see Carayol's report [Ca] (52 pages) or the original paper [HaT].

In a nutshell, let me sketch the basic principle governing the use of geometry. Let X be a (smooth and proper, say) variety over a number field F, and assume X is in some (nice) way acted upon by $\mathrm{GL}_n(\mathbb{A}_F)$. Then on the etale ℓ-adic cohomology spaces of X over an algebraic closure $\overline{F}$ of F, we get two commuting actions of $\mathrm{GL}_n(\mathbb{A}_F)$ and $G_F = \mathrm{Gal}(\overline{F}/F)$. The best hope is that if Σ is an irreducible ℓ-adic representation of G_F and Π an automorphic cuspidal representation of $\mathrm{GL}_n(\mathbb{A}_F)$ such that $\Sigma \otimes \Pi$ "occurs" in such an ℓ-adic cohomology space, then Σ and Π correspond : $L(\Pi_v, s + \frac{k}{2}) = L(\Sigma_v, s)$ for almost all places and some suitable integer k. Of course for that to be true we have to choose X carefully – and be lucky!

9. The model case : modular curves

The model for such a procedure is the case of modular curves $\Gamma\backslash\mathbb{H}$, where $\mathbb{H}$ is the upper half-plane and Γ a congruence subgroup of $SL_2(\mathbb{Z})$. The complex curve $\Gamma\backslash\mathbb{H}$ is the set of complex points of an algebraic curve Y_Γ defined over some number field F; it is smooth if Γ is small enough and can be completed to a proper curve X_Γ/F. The cohomology spaces of X_Γ with constant or more general coefficients can be interpreted in terms of classical modular forms. We can also interpret this way the action of Hecke operators on modular forms. Indeed $\Gamma\backslash\mathbb{H}$ can be seen as $\mathrm{GL}_2(\mathbb{Q})\backslash\mathrm{GL}_2(\mathbb{A}_\mathbb{Q})/\mathbb{R}_+^\times K_\infty K_\Gamma$, where K_∞ is the maximal compact subgroup $0(2)$ of $\mathrm{GL}_2(\mathbb{R})$ and K_Γ is an open subgroup of the product over all prime numbers p of the groups $\mathrm{GL}_2(\mathbb{Z}_p)$. The group $\mathrm{GL}_2(\mathbb{A}_\mathbb{Q})$ does not act on such a quotient space, but there is an action of an algebra of Hecke operators : these are given, for each prime number p, by $\mathrm{GL}_2(\mathbb{Z}_p)$-double classes in $\mathrm{GL}_2(\mathbb{Q}_p)$ and such a class acts by a correspondence on X_Γ. Consequently it also acts on ℓ-adic etale cohomology spaces $H^i(X_\Gamma \otimes \overline{\mathbb{Q}}, \mathbb{Q}_\ell)$. If we take the direct limit $H^i = \varinjlim H^i(X_\Gamma \otimes \overline{\mathbb{Q}}, \overline{\mathbb{Q}}_\ell)$ over smaller and smaller Γ (we choose such $\Gamma's$ with X_Γ defined over $\mathbb{Q}$), we recover on H^i an action of $\mathrm{GL}_2(\mathbb{A}_\mathbb{Q}^\infty)$ (the restricted product of the $\mathrm{GL}_2(\mathbb{Q}_p)$'s for p prime); it is admissible in the sense that each vector has an open stabilizer (smoothness) and open subgroups have only finite dimensional spaces of fixed points. The action of $\mathrm{GL}_2(\mathbb{A}_\mathbb{Q}^\infty)$ on H^i commutes with that of $G_\mathbb{Q}$ and any natural decomposition of H^i with respect to $\mathrm{GL}_2(\mathbb{A}_\mathbb{Q}^\infty)$ will preserve an action of $G_\mathbb{Q}$. In particular, we can isolate (on H^1) the part corresponding to a given weight 2 newform f; specifying the eigenvalues of Hecke operators on f amounts to fixing an irreducible admissible representation π_f^∞ of $\mathrm{GL}_2(\mathbb{A}_\mathbb{Q}^\infty)$ (to get weights > 2 we have to allow non-constant coefficients). The part of H^1 where $\mathrm{GL}_2(\mathbb{A}_\mathbb{Q}^\infty)$ acts via π_f appears as $\pi_f \otimes V$ where V is the space of a 2-dimensional ℓ-adic representation Σ of $G_\mathbb{Q}$ corresponding to

π_f, in the sense above that for almost all prime numbers p, $L(\pi_{f,p}, s-\frac{1}{2}) = L(V_p, s)$. (That last relation can be deduced from the Eichler-Shimura relation).

10. Shimura varieties

If $G = \mathrm{GL}_n$ or more generally if G is any reductive algebraic group over a number field F, we easily get differentiable manifolds analogous to the above modular curves, by forming

$$X_K = G(F)\backslash G(\mathbb{A}_F)/ZK_\infty K^\infty,$$

where K_∞ is a maximal compact subgroup of the Lie group $G(F_\infty)$ (the product over infinite places v of $G(F_v)$), Z the centre of $G(F_\infty)$ and K^∞ is an open compact subgroup of $G(\mathbb{A}_F^\infty)$ (the restricted product over finite places v of the groups $G(F_v)$). It is actually a manifold only if K^∞ is small enough. To each algebraic finite dimensional representation ξ of G, we can attach a local system $\mathcal{L}_\xi$ on X_K and look at the (complex) cohomology spaces $H^i(X_K, \mathcal{L}_\xi \otimes \mathbb{C})$. By relating them to Lie algebra cohomology, the vector spaces $H^i(X_K \mathcal{L}_\xi \otimes \mathbb{C})$ can be interpreted in terms of automorphic forms on G. In particular, a cuspidal automorphic representation Π (the definition of such is similar to the one given above when $G = \mathrm{GL}_n$), with "algebraic" infinity components will usually appear, or rather a corresponding module for a Hecke algebra will appear, in a suitable space $H^i(X_K, \mathcal{L}_\xi)$.

The problem is that such manifolds are not generally algebraic varieties, so there is no way to get ℓ-adic representations. For example if $G = \mathrm{GL}_n$, $n \geq 3$, we do not get algebraic varieties this way. But long ago Shimura singled out a class of G's for which the manifolds X_K indeed have a natural algebraic structure and are defined over number fields. Moreover for most of them, the algebraic structure is obtained because X_K classifies over $\mathbb{C}$ abelian varieties endowed with a number of auxiliary data : this generalizes the case of modular curves, which classify elliptic curves with some extra structure. A similar moduli problem makes sense algebraically and this gives the structure over a number field E (that field is not necessarily equal to F, though). Moreover a variant of the moduli problem usually makes sense, at almost all places v of E, over the ring of integers $\mathcal{O}_{E_v}$ and the result is that X_K has good reduction there, its reduction to the residue field of E_v being again given by a moduli problem through which one can study it.

Again a cuspidal automorphic representation π of $G(\mathbb{A}_F)$, with suitable algebraicity conditions on its infinity part, will usually appear in some ℓ-adic etale cohomology space $H^i(X_K \otimes \overline{E}, \mathcal{L}_\xi \otimes \overline{\mathbb{Q}}_\ell)$, and the hope is to get a corresponding ℓ-adic representation of G_E by looking at the π^∞-isotypic component.

11. Some obstacles

There are many complications to that approach, though, besides the fact that it does not directly apply to GL_n. Firstly X_K as above is not usually proper so that to get a reasonable analysis of the cohomology, one would have to construct a smooth equivariant compactification, with possibly the boundary again given by moduli problems ... Secondly it is not so straightforward to analyze the cohomology spaces in terms of automorphic representations. Langlands has shown that in such an analysis, automorphic representations of groups other than G sneak in : this is the phenomenon of endoscopy. Thirdly it is not entirely clear where and how often a given cuspidal automorphic representation of G over F appears in cohomology and indeed such a multiplicity problem is responsible for the occurrence of the integer a in Theorem 1 of Lecture 1.

Moreover the technical complications are enormous. The standard approach, pioneered by Langlands, is to analyze the geometry of X_K via the Lefschetz-Grothendieck trace formula and the space of automorphic forms via the Arthur-Selberg trace formula. Two formidable tools, which have to be related to each other ...

12. Kottwitz's "simple" Shimura varieties

However R. Kottwitz singled out some "simple" Shimura varieties where most of the problems above can be solved or circumvented, and those varieties are used by Harris and Taylor. They are attached to unitary groups with respect to a quadratic extension F/F^+ of number fields, where F^+ is totally real and F totally imaginary. Let me describe the ones Harris and Taylor use which are "simple" in Kottwitz's sense. One chooses a division algebra B with center F and degree n^2 over F, with a positive involution of second kind $*$ (that is the restriction of $*$ to F is the non-trivial involution of F over F^+). There are other technical hypotheses, but the important requirement to get local results for our given p-adic field F_0, is the following : F_0 appears as a completion of F^+, and F and B are split at that place 0, so that F has two completions over F_0^+, one of which is F_0, and $B_0 = B \otimes_F F_0$ is isomorphic to $M_n(F_0)$, $B_0^\times$ to $\mathrm{GL}_n(F_0)$. At least for the given p-adic field F_0, we recover from $B^\times$ the group we want!

Conjugating the involution $*$ by a $*$-antisymmetric element β, we get another involution $\natural$ and a reductive group G over $\mathbb{Q}$ defined by

$$G(R) = \{g \in B^{op} \otimes R, bb^\natural \in R^\times\} \text{ for any } \mathbb{Q}-\text{algebra } R\,.$$

It is a form of the group of unitary similitudes of degree n over F. The element β is chosen so that group has signature $(n-1, 1)$ at one infinite place of F, and $(n, 0)$ at the other infinite places. If K^∞ is a compact open subgroup of $G(\mathbb{A}_\mathbb{Q}^\infty)$, we can form X_K as above, which is a compact manifold if K^∞ is small enough : it is in fact the union of a finite number of quotients $\Gamma\backslash\mathcal{X}$ where $\mathcal{X}$ is the unit ball in $\mathbb{C}^{n-1}$ and Γ a discrete cocompact subgroup of $G(\mathbb{R})$. In this case X_K is a proper algebraic variety over some number field, given by a moduli problem : it classifies abelian varieties up to isogeny, of dimension $[F^+ : \mathbb{Q}]n^2$, with an action of B (and, of course, other auxiliary data and compatibility requirements). Besides the group G presents no endoscopy problem. But, to expect Galois representations of G_F, it is better if X_K is defined over F! One way to obtain that is to assume $F = F^+E$, where E is a quadratic imaginary field (where p splits). We shall assume this from now on.

13. Results of Kottwitz

At almost all prime numbers p, X_K has good reduction, and a model for the reduction is given by a moduli problem classifying abelian varieties with an action of an order O_B in B, plus other data of course. Kottwitz analyzed enough of $X_K(k)$

for finite fields k to obtain, via the Grothendieck-Lefschetz and the Arthur-Selberg trace formulae, the following result.

Let Π be an automorphic cuspidal representation of $G(\mathbb{A}_Q)$ with algebraicity (and regularity) conditions on its infinity component Π_∞, and a mild requirement at some finite place. Then, for some algebraic finite dimensional representation ξ of G, the finite part Π^∞ of Π "occurs" in the etale ℓ-adic cohomology space $H^{n-1} = \varinjlim(X_K \times \overline{F}, \mathcal{L}_\xi \otimes \overline{\mathbb{Q}}_\ell)$. The space of $G(\mathbb{A}_{\mathbb{Q}}^\infty)$-embeddings of Π^∞ into H^{n-1} has finite dimension $a \cdot n$ for some positive integer a, and if Σ' is the action of G_F on that space, we have, at almost all places v of F

$$L(\Pi_v, s - \frac{n-1}{2}) = L(\Sigma'_v, s)^a \ .$$

(As for GL_n, Π is unramified at almost all finite places v of F; in fact for almost all prime numbers r, there is a non-trivial fixed vector under a compact subgroup of $G(\mathbb{Q}_r)$ of maximal volume; at places v of F above such r, Π_v has an L-factor of the form $P(q_v^{-s})^{-1}$, $P \in \mathbb{C}[X]$, $P(0) = 1$).

We are now close to Thm. 1, except that Π is an automorphic cuspidal representation of G, not of GL_n over F!

14. Results of Clozel

However, if we base change G from $\mathbb{Q}$ to the quadratic imaginary field E, we get a group close to the algebraic group defined by $B^{op\times}$, itself an interior form of GL_n/F. Such a close relationship between groups reflects on the side of automorphic representations : this is Langlands' theory of base change, or more generally Langlands' functoriality.

The situation at hand was explored by Clozel, and the result is roughly the following : the automorphic cuspidal representation Π of $G(\mathbb{A}_{\mathbb{Q}})$ above "base changes" to an automorphic cuspidal representation Π' of $\mathrm{GL}_n(\mathbb{A}_F)$ such that :
(1) $\Pi'^\vee = \Pi'^c$ where c is the complex conjugation of F over F^+;
(2) for almost all places v of F, $L(\Pi'_v, s) = L(\Pi_v, s)$. Conversely a cuspidal automorphic representation Π' of $\mathrm{GL}_n(\mathbb{A}_F)$, satisfying (1) and the other conditions of Thm. 1, is such a base change. This gives a result slightly weaker than Thm. 1, at least in the case where $F = F^+E$, E quadratic imaginary.

15. Results of Harris and Taylor

After receiving the first version of [HT], I realized that using [Ha 2], the above case of Thm. 1 is enough to get the Langlands correspondence as stated in Thm. 2 [He]. But Harris and Taylor do much more : they actually get a geometric model for that local correspondence, and in turn this has global consequences as Thm. 3. That feat actually takes most of their manuscript, and I shall now give a (too) brief description of it. Let me mention though that, in doing so, they complete a program initiated by Carayol.

Most of the work consists in analyzing the Shimura variety X_K at finite places w of bad reduction (actually, Harris and Taylor also essentially reprove in their case Kottwitz's result for good reduction, by a different method).

At such a place w, X_K has an integral model over O_{F_w}, which is projective but not smooth, and is again given by a moduli problem classifying abelian varieties with an action of O_B. The special fiber has a filtration by closed subvarieties, indexed by a measure of "supersingularity" : this is analogous to the consideration of supersingular versus ordinary elliptic curves on the special fiber of modular curves. The most supersingular points form a finite subset of X_K. The etale ℓ-adic cohomology groups $H^i(X_K \otimes \overline{F}_w, \mathcal{L}_\xi \otimes \overline{\mathbb{Q}}_\ell)$ are studied via vanishing cycles on the special fiber : there is a spectral sequence of vanishing cycles

$$H^i(X_K \otimes \overline{k}_w, R\Phi^j \otimes \mathcal{L}_\xi \otimes \overline{\mathbb{Q}}_\ell) \Longrightarrow H^{i+j}(X_K \otimes \overline{F}_w, \mathcal{L}_\xi \otimes \overline{\mathbb{Q}}_\ell),$$

where the $R\Phi^j$'s are sheaves of vanishing cycles. They are analyzed stratum by stratum.

The main point (but for lack of time and expertise, I shall refrain from giving details here) is that the sheaves $R\Phi^j$ are related to purely local objects, which are spaces of ℓ-adic vanishing cycles attached to some specially constructed formal schemes. My last section will describe such formal schemes.

Remark. Berkovich's theory of ℓ-adic cohomology for formal schemes is absolutely crucial here. In fact Harris and Taylor needed a bit more than previously known, and that was provided by Berkovich : see the appendix of [Ha T].

16. Formal modules and Drinfeld's moduli varieties

Here the base field is our p-adic field F_0. Let $\mathcal{O}$ be its ring of integers, and ϖ a uniformizer of F_0. Fix the integer $n \geq 1$ and the prime number $\ell \neq p$.

Formal $\mathcal{O}$-modules, i.e. formal groups with an action of $\mathcal{O}$, have been used by Lubin and Tate to give a purely local proof of local class field theory. Drinfeld, using deformations of such objects, has constructed formal schemes with commuting actions of $\mathrm{GL}_n(F_0)$ and W_{F_0}. Berkovich's spaces of ℓ-adic vanishing cycles on such formal schemes then give commuting actions of $\mathrm{GL}_n(F_0)$ and W_{F_0} on ℓ-adic vector spaces, which implement the Langlands local correspondence. Let us be a little more precise.

We consider formal $\mathcal{O}$-modules of fixed height n (the height controls how ϖ acts in characteristic p). Over the algebraic closure $\overline{k}$ of the residue field of $\mathcal{O}$ there is a unique such formal module Σ up to isomorphism. Its endomorphism ring is the ring of integers O_D of the division algebra D with center F_0, degree n^2 over F_0 and invariant $1/n$ in the Brauer group of F_0. The functor of deformations of Σ is prorepresentable by a local complete algebra R, isomorphic in fact to $\widehat{\mathcal{O}}[[T_1, \dots T_{n-1}]]$, where $\widehat{\mathcal{O}}$ is the ring of integers of the completion of the maximal unramified extension of F_0 in $\overline{F}_0$. On the formal scheme $Spf(R)$, Σ has a universal deformation $\widetilde{\Sigma}$. As for elliptic curves, there is a notion of level m structure for such formal $\mathcal{O}$-modules. On the generic fiber it is simply an isomorphism of the ϖ^m-torsion points of $\widetilde{\Sigma}$ with the $\mathcal{O}$-module $(\varpi^{-m}\mathcal{O}/\mathcal{O})^h$. The general notion is due to Drinfeld. The functor classifying such level m-structures is prorepresentable by a local complete algebra R_m. On $Spf(R_m)$ the group $\mathrm{GL}_n(\mathcal{O}/\varpi^m\mathcal{O})$ acts.

17. Group actions on the vanishing cycles

Berkovich's theory of ℓ-adic cohomology for p-adic analytic spaces attaches vanishing cycle spaces Ψ^i_m to $Spf(R_m)$. Varying m, we get an inductive system of ℓ-adic vector spaces; write Ψ^i for the limit (in fact the union). On Ψ^i we have commuting actions of $\mathcal{O}_D^\times$ (by automorphisms on the base), of $\mathrm{GL}_n(\mathcal{O})$ (by automorphisms of the level structure), and of the inertia group I_{F_0}. From Ψ^i one constructs a slightly larger space $\widetilde{\Psi}^i$ with more group action : on $\widetilde{\Psi}^i$ we have an **admissible** action of $\mathcal{O}_D^\times \times \mathrm{GL}_n(F_0) \times W_{F_0}$. If τ is a given irreducible admissible representation of $\mathcal{O}_D^\times$ (such a τ is finite dimensional), we let $\widetilde{\Psi}^i(\tau)$ be the representation of $\mathrm{GL}_n(F_0) \times W_{F_0}$ on the τ-isotypic component of $\widetilde{\Psi}^i$. It is admissible of finite length and we put

$$\Psi^*(\tau) = (-1)^{n-1}\Sigma(-1)^i[\widetilde{\Psi}^i(\tau)]$$

where [ ] denotes a class in the Grothendieck group of (virtual) admissible finite length representations of $\mathrm{GL}_n(F_0) \times W_{F_0}$. The following theorem describes geometrically the local Langlands correspondence $\pi \mapsto \sigma(\pi)$ of Part A I), at least in the crucial case where π is supercuspidal.

THEOREM (Harris-Taylor). *Assume π is supercuspidal. Then there is a unique irreducible admissible representation τ of $\mathcal{O}_D^\times$ such that* $\Psi^*(\tau) = [\pi \otimes (\sigma(\pi^\vee)| |^{\frac{n-1}{2}})]$.

(Here $|\ |$ is the norm character of W_{F_0}).

In fact τ is obtained from π by the Jacquet-Langlands correspondence, which is an instance of Langlands' functoriality between the local groups $\mathrm{GL}_n(F_0)$ and $D^\times$.

B) A local-global principle

In this part B, I want to illustrate several instances of the local-global principle already mentioned in part A, and in particular I want to show how it applies to give a proof of the Langlands conjectures for GL_n over p-adic fields, following [Ha2, HaT, He]. I shall assume some familiarity with number fields, their completions, and their adele rings and idele groups.

B.1 Class-field theory

B.1.1 Class-field theory describes all finite abelian extensions of numbers fields, i.e. finite extensions of $\mathbb{Q}$, and p-adic fields, i.e. finite extensions of the field $\mathbb{Q}_p$ of p-adic numbers. This is an elaborate theory, its exposition requires books. I shall only give the salient facts refering to e.g., [Lg 1, Ne, We] for details. In the sequel, p is a prime number, F a number field, and K a p-adic field. We shall also refer here to number field as **global** fields and to p-adic fields as **local** fields.

All finite abelian extensions of $\mathbb{Q}$ or $\mathbb{Q}_p$ are cyclotomic, i.e. are subextensions of fields obtained by adjoining some roots of unity; this is the celebrated theorem of Kronecker-Weber. For general F and K the description is more complicated. For a number field F, it involves the adele ring $\mathbb{A}_F$ of F and its idele class group $F^\times\backslash\mathbb{A}_F^\times$ (it can also be expressed more classically in terms of ideals of the ring of integers $\mathcal{O}_F$ of F cf. [Lg 1, Jz].)

The adele ring $\mathbb{A}_F$ of F is defined in terms of the completions F_v of F at the different places v of F. There are the *infinite places* of F, the real ones which correspond to the different field embeddings of F into $\mathbb{R}$, and the complex ones which correspond to pairs of complex conjugate field embeddings of F into $\mathbb{C}$ as a dense subfield. On the other hand, there are the *finite* places of F, one for each maximal ideal v of the ring of integers $\mathcal{O}_F$ of F; if ℓ is the residue characteristic of v, i.e. the characteristic of the finite field $k_v = \mathcal{O}_E/v$, then the corresponding completion F_v is an ℓ-adic field; when $F = \mathbb{Q}$ and v is given by a prime number ℓ in $\mathbb{Z} = \mathcal{O}_\mathbb{Q}$, then the completion is none other than the field of ℓ-adic numbers $\mathbb{Q}_\ell$.

It is therefore natural to consider, alongside the number fields, also the p-adic fields for prime numbers p (and also the archimedean complete fields $\mathbb{R}$ and $\mathbb{C}$, but their finite abelian extensions are easy to describe). Such a p-adic field K has a prefered absolute value $|\ \ |_K$; its ring of integers $\mathcal{O}_K = \{x \in K \mid |x|_K \leq 1\}$ has a unique maximal ideal $P_K = \{x \in K \mid |x|_K < 1\}$, and the residue field $k_K = \mathcal{O}_K/P_K$ is a finite extension of $\mathbb{F}_p = \mathbb{Z}/p\mathbb{Z}$; the ideal P_K is principal and its generators are called **uniformizers** of K. When $K = \mathbb{Q}_p$ we have $\mathcal{O}_K = \mathbb{Z}_p$, $P_K = p\mathbb{Z}_p$. A given p-adic field K appears as a completion F_v of some number field F at a finite place v : that is to say, there is a choice of F and v such that F_v is topologically isomorphic to K. In fact there are very many different choices for F and v.

B.1.2 The adele ring $\mathbb{A}_F$ of F is defined as a union of subrings $\mathbb{A}_F^{(S)} = \Pi_{v\in S}F_v\Pi_{v\notin S}$ $\mathcal{O}_{F_v}$, where S runs through finite sets of places of F containing the infinite places. It is a topological ring if we put the product topology on $\mathbb{A}_F^{(S)}$ and the inductive limit topology on $\mathbb{A}_F$. The field F then sits diagonally in $\mathbb{A}_F$ as a discrete subring, although of course, it is dense in each of its completions F_v. Similarly the *idele group* $\mathbb{A}_F^\times$ of units of $\mathbb{A}_F$ is a union of subgroups $\mathbb{A}_F^{\times(S)} = \Pi_{v\in S}F_v^\times\Pi_{v\notin S}\mathcal{O}_{F_v}^\times$; it is a topological group, again, if we put the product topology on $\mathbb{A}_F^{\times(S)}$ and the inductive limit topology on $\mathbb{A}_F^\times$, which turns $\mathbb{A}_F^\times$ into a locally compact topological group. This is not the topology on $\mathbb{A}_F^\times$ induced by the one on $\mathbb{A}_F$.

As before $F^\times$ sits diagonally in $\mathbb{A}_F^\times$ as a discrete subgroup. Also for each place v of F, $F_v^\times$ sits inside $\mathbb{A}_F^\times$, as the v-component. The quotient $C_F = \mathbb{A}_F^\times / F^\times$ is called the *idele class group* of F : When F is $\mathbb{Q}$, $\mathbb{A}_\mathbb{Q}^\times$ appears as the direct product of its diagonal subgroup $\mathbb{Q}^\times$ and the subgroup $\mathbb{R}_+^\times \Pi_{\ell\,\text{prime}} \mathbb{Z}_\ell^\times$, where $\mathbb{R}_+^\times$ is the group of positive real numbers.

B.1.3 Let E/F be a finite extension of number fields. The norm homomorphism $N_{E/F}$ from $E^\times$ to $F^\times$ extends naturally to a norm homomorphism $N_{E/F}$ from $\mathbb{A}_E^\times$ to $\mathbb{A}_F^\times$. More precisely let v be a place of F; then there are finitely many places w of F *above* v (one writes $w|v$) : if v is infinite, this means the embeddings of E into $\mathbb{R}$ or $\mathbb{C}$ defining w extend those defining v, whereas if v is finite, this means the maximal ideal w of $\mathcal{O}_E$ contains v. The tensor product $F_v \otimes_F E$ splits naturally as a product of the finite extensions E_w/F_v where w runs through the places of E above v. The natural norm on the F_v-algebra $F_v \otimes_F E$ induces on each extension E_w the norm N_{E_w/F_v}, and the norm from $\mathbb{A}_E^\times$ to $\mathbb{A}_F^\times$ coincides with N_{E_w/F_v} on $E_w^\times$ and with $N_{E/F}$ on the diagonal $E^\times$.

We still write $N_{E/F}$ fo the induced norm map from C_E to C_F.

The main result of class-field theory for number fields is the following.

THEOREM 1. — *To each finite abelian extension E of the number field F associate its norm group $N_E = N_{E/F}(C_E)$. Then N_E is an open subgroup of finite index of C_F, and there is a canonical group isomorphism (reciprocity map) τ_E : $C_E/N_E \simeq \mathrm{Gal}(E/F)$. Moreover the map $E \mapsto N_E$ induces a bijection between finite abelian extensions E of F, up to isomorphism, and open subgroups of finite index of C_F.*

B.1.4 There is a similar result for p-adic fields, where the role of the idele class group is played by the multiplicative group.

THEOREM 2. — *To each finite abelian extension L of the p-adic field K associate its norm group $N_L = N_{L/K}(L^\times)$. Then N_L is an open subgroup fo finite index of $K^\times$, and there is a canonical group isomorphism (reciprocity map) τ_L : $K^\times/N_L \simeq \mathrm{Gal}(L/K)$. Moreover the map $L \mapsto N_L$ induces a bijection between finite abelian extensions L of K, up to isomorphism, and open subgroups of finite index of $K^\times$.*

As is to be expected the theorem for p-adic fields just stated and the theorem for number fields above are strongly related. Let me explain that now.

Let E be a finite *Galois* extension of the number field F, and let v be a place of F, w a place of E above v. Then E_w is Galois over F_v, and , as each F_v-automorphism of E_w induces an F-automorphism of E, we get an injective group homomorphism of $\mathrm{Gal}(E_w/F_v)$ into $\mathrm{Gal}(E/F)$. All the different places w of E above a given place v of F form only one orbit under the action of $\mathrm{Gal}(E/F)$, hence the subgroups $\mathrm{Gal}(E_w/F_v)$ are conjugate to each other; in particular, when E is abelian over F, they are all equal to the same group, called the decomposition subgroup at v.

The compatibility between the results above for number fields and for p-adic fields can be expressed in the following manner

THEOREM 3. — *Let v be a finite place of the number field F. Let E be a finite abelian extension of F and w a place of E above v. Then the following diagram is commutative*

$$\begin{array}{ccc} F_v^\times/N_{E_w/F_v}(E_w^\times) & \xrightarrow{\tau_{E_w}} & \mathrm{Gal}(E_w/F_v) \\ \downarrow & & \downarrow \\ C_F/N_{E/F}(C_E) & \xrightarrow{\tau_E} & \mathrm{Gal}(E/F), \end{array}$$

where the left-hand vertical map is induced by the embedding $F_v^\times \to \mathbb{A}_F^\times$, and the right-hand vertical map is the canonical inclusion of the decomposition group into $\mathrm{Gal}(E/F)$.

In fact, the exact same statement is true for an infinite place v. Also, the maps τ_{E_w} are the same for all places w of E above v.

B.1.5 Nowadays (see [We], [Ne]) class-field theory is first established for local fields, i.e. Theorem 2 is proved first. Then a homomorphism $\tilde{\tau}_E : \mathbb{A}_F^\times \to \mathrm{Gal}(F/F)$ is constructed out the local maps τ_{E_w}, and Theorem 1 is proved, so that Theorem 3 holds essentially by definition. The main step there is to prove that the map $\tilde{\tau}_E$ is trivial on the diagonal subgroup $F^\times$ of $\mathbb{A}_F^\times$.

However the classical approach (see [Lg 1]) is different. One proves first Theorem 2 in the easy case where the extension L/K is *unramified*, in which case $N_{L/K}(L^\times)$ contains the unit group $\mathcal{O}_K^\times$ (see below). If E is a fixed abelian extension of F, then for all finite places w of E except a finite number (it is customary to say "*for almost all*" places w of E), the corresponding extension E_w/F_v of local fields is unramified. In this way one can define a homomorphism $\tilde{\tau}_E$ into $\mathrm{Gal}(E/F)$ but only on those elements of $\mathbb{A}_F^\times$ which have trivial component at the ramified (i.e. not unramified) places : indeed, for such an element $x = (x_v)$ of $\mathbb{A}_F^\times$, $\tau_{E_w}(x_v) = y_v$ is defined (because E_w/F_v is unramified) and is trivial for almost all v since x_v is a unit in $\mathcal{O}_{F_v}$ for almost all v, one then defines $\tilde{\tau}_E(x)$ to be the product of the y_v's.

The most serious step is then to prove that this partially defined homomorphism $\tilde{\tau}_E$ *can be extended* to a (continuous) homomorphism of $\mathbb{A}_F^\times$ into $\mathrm{Gal}(E/F)$ which is *trivial* on $F^\times$. This then defines a (continuous) homomorphism τ_E of C_F into $\mathrm{Gal}(E/F)$ and one shows Theorem 1.

Then it is possible to deduce Theorem 2 when the extension L/K is possibly ramified. Firstly, in the situation of Theorem 1, if v is a place of F and w a place of E above v (possibly ramified) one has to show that the image of $F_v^\times$ under τ_E is the subgroup $\mathrm{Gal}(E_w/F_v)$ of $\mathrm{Gal}(E/F)$, and τ_E does induce an isomorphism of $F_v^\times/N_{E_w/F_v}(E_w^\times)$ onto $\mathrm{Gal}(E_w/F_v)$. Secondly if K is a given p-adic field, and L a finite abelian extension of K, one can easily find a number field F, a finite abelian extension E of F, a place v of F and a place w of E above v, such that the extensions E_w/F_v and L/K are (topologically) isomorphic. By transport of structure, one then gets an isomorphism of $K^\times/N_{L/K}(L^\times)$ onto $\mathrm{Gal}(L/K)$. The final step is to show that this last isomorphism depends only the local situation L/K and not on the global situation E/F, $w|v$ we have embedded it in (and, of course, that we recover the right map when L/K is unramified).

So the classical approach would appear much more complicated than the modern one. However the principle behind it, which I like to call a local-global principle, can be applied in situations when there is no direct construction of the local theory, or no direct way to study the local situation. We shall see later instances of that principle.

B.1.6 But before we pursue, I want to explain what Theorem 2 means explicitely when L/K is unramified; and also what Theorem 1 says when F is $\mathbb{Q}$, where many abelian extensions (indeed all) are known, viz. the cyclotomic fields. In fact, it is important to point out that the knowledge of cyclotomic extensions is an essential ingredient in passing from Theorem 2 —at all places v of a number field or simply at the finite places which are unramified in a given abelian extension— to Theorem 1. However special the information appears to be, it is crucial in such local-global arguments to have, at least in *some* global situations, a precise control.

Let us first explain the unramified case of Theorem 2, then. A finite extension L of the p-adic field K is called *unramified* if the residue field degree $[k_L : k_K]$ is equal to the degree $[L : K]$; equivalently, if a uniformizer for K is also a uniformizer for L. Then L is obtained from K by adjoining the roots of $X^{|k_L|} - X$. The extension L/K is Galois : the Galois group $\mathrm{Gal}(L/K)$ acts naturally on the residue field k_L by k_K-automorphisms and, for L/K unramified, the induced morphism of $\mathrm{Gal}(L/K)$ into $\mathrm{Gal}(k_L/k_K)$ is an isomorphism; in particular L is cyclic over K and $\mathrm{Gal}(L/K)$ has a prefered generator, the arithmetic Frobenius element, which acts on the residue field as $x \mapsto x^{|k_K|}$.

On the other hand, every element from $\mathcal{O}_K^\times$ is a norm from $\mathcal{O}_L^\times$, and since uniformizers of K are also uniformizers of L, the group $K^\times/N_{L/K}(L^\times)$ is also cyclic of order $[L : K]$, which a prefered generator given by the class of uniformizers.

Class field theory for the unramified extension L/K is the simple statement that we have a canonical isomorphism

$$\tau_L : K^\times/N_{L/K}(L^\times) \simeq \mathrm{Gal}(L/K),$$

since both sides are cyclic of the same degree with a prefered generator! It is now customary to normalize the isomorphism so that the uniformizers of K, or rather their class in $K^\times/N_{L/K}(L^\times)$, correspond to the *geometric Frobenius elements* which are the inverses the arithmetic Frobenius elements.

Remark : In the case of (possibly ramified) extensions L/K of p-adic fields, the reciprocity maps τ of Theorem 2 can be characterized by what happens in the unramified case, plus some natural compatibilities when L varies. See [Ne] for such an approach.

B.1.7 Let us now explain the content of Theorem 1 when F is $\mathbb{Q}$, and E is the cyclotomic field $\mathbb{Q}(\chi)$ generated by a primitive N-th root of unity, N being a fixed integer, $N \geq 2$. In that case, the Galois group $\mathrm{Gal}(\mathbb{Q}(\chi)/\mathbb{Q})$ can be identified with $(\mathbb{Z}/N\mathbb{Z})^\times$ of an integer a prime to N acting as the automorphism of $\mathbb{Q}(\chi)$ sending χ to χ^a. The finite places of $\mathbb{Q}$ correspond to the prime numbers. All prime numbers p not dividing N are unramified in $\mathbb{Q}(\chi)$ (i.e. all places w of $\mathbb{Q}(\chi)$ above p have

$\mathbb{Q}(\chi)_w/\mathbb{Q}_p$ unramified), and the corresponding arithmetic Frobenius element acts as $\chi \mapsto \chi^p$ on χ, so is identified with the class of p in $(\mathbb{Z}/N\mathbb{Z})^\times$.
It is easy to verify that there is a unique homomorphism

$$\tilde{\tau} : \mathbb{A}_{\mathbb{Q}}^\times \to (\mathbb{Z}/N\mathbb{Z})^\wedge$$

which is

- trivial on the diagonal $\mathbb{Q}^\times$
- trivial on the factor $\mathbb{R}_+^\times$ at the infinite place,
- for a prime p dividing N, trivial on elements of $\mathbb{Q}_p^\times$ congruent to 1 mod $N\mathbb{Z}_p$
- for a prime p not dividing N, trivial on $\mathbb{Z}_p^\times$ and sending the uniformizer p to the class of p^{-1} in $(\mathbb{Z}/N\mathbb{Z})^\times$. Since $\tilde{\tau}$ is trivial on $\mathbb{Q}^\times$ it induces a homomorphism $C_{\mathbb{Q}} \to (\mathbb{Z}/N\mathbb{Z})^\times$. One shows it is surjective (*Čebotarev's theorem*) and that its kernel is $N_{\mathbb{Q}(\chi)/\mathbb{Q}}(C_{\mathbb{Q}(\chi)})$, hence providing the reciprocity map

$$\tau_{\mathbb{Q}(\chi)} : C_{\mathbb{Q}}/N_{\mathbb{Q}(\chi)} \simeq \mathrm{Gal}(\mathbb{Q}(\chi)/\mathbb{Q}) = (\mathbb{Z}/N\mathbb{Z})^\times.$$

B.1.8 The reciprocity maps have some natural properties. An important one is when we consider a subextension of an abelian extension. In the global case, if E is a finite abelian extension of the number field F, and E' a subextension, the following diagram is commutative

$$\begin{array}{ccc} C_F/N_E & \xrightarrow[\approx]{\tau_E} & \mathrm{Gal}(E/F) \\ \downarrow & & \downarrow \\ C_F/N_{E'} & \xrightarrow[\approx]{\tau_{E'}} & \mathrm{Gal}(E'/F) \end{array}$$

where the left vertical map comes from the obvious inclusion $N_E \subset N_{E'}$ and the right vertical map is the canonical quotient obtained by restricting to E' the F-automorphisms of E. We have a similar commutativity statement in the local case.

This makes it possible to consider all finite abelian xtensions E of F at the same time. Let us fix an algebraic closure $\overline{F}$ of F. On the Galois group $\mathrm{Gal}(\overline{F}/F)$ of F-automorphisms of F put the topology of pointwise convergence on $\overline{F}$; then $\mathrm{Gal}(F/F)$ is a compact topological group, and appears as a profinite group, moro precisely, the projective limit of its finite (discrete) quotients $\mathrm{Gal}(E/F)$, where E runs through finite Galois extensions of F in $\overline{F}$.

One can restate the main theorem of global class field theory by saying that there is a canonical isomorphism of topological groups of the abelianized Galois group $\mathrm{Gal}^{\mathrm{ab}}(\overline{F}/F)$ (i.e. the quotient of $\mathrm{Gal}(\overline{F}/F)$ by the closure of its commutator group) and the quotient of C_F by some closed subgroup D_F, which actually turns out to be its connected component (see [AT, ch. 9]).

Similarly if $\overline{K}$ is an algebraic closure of the p-adic K, the abelianized Galois group $\mathrm{Gal}^{\mathrm{ab}}(\overline{K}/K)$ is identified with the completion of $K^\times$ with respect to the topology of finite index subgroups.

B.1.9 However, it is more fruitful to think in dual terms, going to the Fourier transform side. Global class field theory says that you have a canonical identification of quasicharacters [1] of $\mathrm{Gal}(\overline{F}/F)$ and finite order characters of C_F. Indeed, because $\mathrm{Gal}(\overline{F}/F)$ is profinite, a quasicharacter χ of $\mathrm{Gal}(\overline{F}/F)$ factorizes through a finite abelian quotient of $\mathrm{Gal}(E/F)$ hence corresponds via τ_E to a finite order character of $C_F/N_{E/F}(C_E)$. By (1.8.1) the induced character π_χ of C_F does not depend on the choice of E. Similarly local class field theory identifies quasicharacters of $\mathrm{Gal}(\overline{K}/K)$ (necessarily of finite order) and finite order characters of $K^\times$.

I shall now make those identifications more explicit since the Langlands conjectures suggest a generalization, involving not only quasicharacters of the Galois group i.e. representations of dimension 1, but representations of higher dimension.

If K is our p-adic field, the Galois group $\mathrm{Gal}(\overline{K}/K) = G_K$ has a canonical closed subgroup, the inertia group I_K , which is the subgroup fixing the union K^{nr} of the finite unramified extensions of K in $\overline{K}$. The quotient G_K/I_K is a topological group free with one topological generator Φ_K (the Geometric Frobenius element) acting as $x \mapsto x^{1/|k_K|}$ on the residue field of K^{nr}. If a quasicharacter χ of $\mathrm{Gal}(\overline{K}/K) = G_K$ is *unramified*, i.e. if χ is trivial on I_K, then the corresponding quasicharacter $\pi(\chi)$ of $K^\times$ is also *unramified* i.e. trivial on $U_K^\times$ and its value an a uniformizer $\widetilde{\omega}_K$ is $\pi(\chi)(\widetilde{\omega}_K) = \chi(\Phi_K)$.

If now F is our number field, and v a place of F, then any extension of v to the algebraic closure $\overline{F}$ of F yields an algebraic closure $\overline{F}_v$ of F_v and an embedding of $\mathrm{Gal}(\overline{F}_v/F_v)$ as a topological subgroup of $\mathrm{Gal}(\overline{F}/F)$. (The choice of another extension of v yields a conjugate embedding). If χ is a quasicharacter of $\mathrm{Gal}(\overline{F}/F)$ then it is *unramified almost everywhere*, meaning that its restriction χ_v to $\mathrm{Gal}(\overline{F}_v/F_v) = G_{F_v}$ is, for almost every finite v of F, trivial on the inertia group I_{F_v}, hence is characterized by its value on the Frobenius element $\chi_v(\Phi_{F_v})$.

Similarly, a quasicharacter λ of C_F (or, which amounts to the same but is more customary, a quasicharacter λ of $\mathbb{A}_F^\times$ trivial on $F^\times$) is *unramified almost everywhere*, meaning that for almost all finite places v of F, the restriction λ_v of λ to the subgroup $F_v^\times$ of $\mathbb{A}_F^\times$ is unramified.

If χ is a quasicharacter of $\mathrm{Gal}(\overline{F}/F)$ then the quasicharacter $\pi(\chi)$ attached to χ by class-field theory is the **unique one** such that, for every place v of F where χ_v is unramified, $\pi(\chi)_v$ is also unramified and verifies $\pi(\chi)_v(\varpi_{F_v}) = \chi_v(\Phi_{F_v})$ at uniformizers ϖ_{F_v} of F_v.

Of course then, at **any** place v of F, the quasicharacter $\pi(\chi)_v$ of $F_v^\times$ is associated by local class-field theory to the quasicharacter χ_v of $\mathrm{Gal}(\overline{F}_v/F_v)$.

B.1.10 R.P. Langlands advocated, to try and understand better the structure of $\mathrm{Gal}(\overline{F}/F)$, to look at higher dimensional representations, and associate them with automorphic representations of the groups $\mathrm{GL}_n(\mathbb{A}_F)$. It is our task in the following chapters to describe the conjectures (as before, there are variants over number fields and p-adic fields, strongly interrelated) and to describe also the local-global principle which has led Harris and Taylor ([Ha 2], [HT], see also [He]) to a complete proof of the conjectures for p-adic fields.

[1](For us a quasicharacter of a topological group is a continuous homomorphism into $\mathbb{C}^\times$; it is a character if its values have absolute value 1. A finite order quasicharacter is thus a character.)

B.2 Langlands' conjectures for $\mathrm{GL}(n)$ over number fields

B.2.1 So the idea is to try and understand better the Galois group $\mathrm{Gal}(\overline{F}/F)$ by investigating its representations on finite dimensional complex vector spaces. For a topological group like $\mathrm{Gal}(\overline{F}/F)$, the natural notion is that of continuous representations. Because $\mathrm{Gal}(\overline{F}/F)$ is profinite and $\mathrm{GL}_n(\mathbb{C})$ has a neighbourhood of 1 containing no non-trivial subgroup, a **continuous** representation $\sigma : \mathrm{Gal}(\overline{F}/F) \to \mathrm{Aut}_{\mathbb{C}}(V)$ on a finite dimensional complex vector space V is the same as a representation which factorizes through a **finite** discrete quotient $\mathrm{Gal}(E/F)$, where E is a finite Galois extension of F in $\overline{F}$. So in effect we are looking at all finite Galois extensions E of F, and the (usual) complex representations of $\mathrm{Gal}(E/F)$.

Note that there is a natural obvious notion of isomorphism for such representations of $\mathrm{Gal}(\overline{F}/F)$, and the isomorphism classes of complex 1-dimensional representations of $\mathrm{Gal}(\overline{F}/F)$ correspond exactly to the quasicharacters $\mathrm{Gal}(\overline{F}/F) \to \mathbb{C}^\times$.

B.2.2 However there are other, arguably more important, sources of representations of $\mathrm{Gal}(\overline{F}/F)$. Indeed, consider an algebraic variety X over F, proper and smooth, say. If ℓ is a prime number, then the machinery of étale ℓ-adic cohomology gives finite-dimensional vector spaces over $\mathbb{Q}_\ell$, $H_\ell = H^i(X(\overline{F}), \mathbb{Q}_\ell)$, for integers i between 0 and $2\dim(X)$. On those vector spaces, the group $\mathrm{Gal}(\overline{F}/F)$ acts via its action on $X(\overline{F})$, and we get continuous homomorphisms ("ℓ-adic representations") $\mathrm{Gal}(\overline{F}/F) \to \mathrm{Aut}_{\mathbb{Q}_\ell}(H_\ell^i)$.

Variants allow to put coefficient systems on X, to consider more general varieties or even motives [Mot], or to extend the coefficient-field $\mathbb{Q}_\ell$ to finite extensions or to an algebraic closure $\overline{\mathbb{Q}}_\ell$.

Such ℓ-adic representations usually have **infinite** image.

Example. Assume $F = \mathbb{Q}$ and take for X an elliptic curve E over $\mathbb{Q}$. In particular $E(\overline{\mathbb{Q}})$ is an abelian group. For each positive integer r we can consider the subgroup $E[\ell^r]$ of $E(\overline{\mathbb{Q}})$ made out of points killed by ℓ^r. As r varies, they form a projective limit $T_\ell(E)$, the ℓ-adic Tate module, which is a $\mathbb{Z}_\ell$-free module of rank 2. The $\mathbb{Q}_\ell$-vector space $V_\ell(E) = T_\ell(E) \otimes_{\mathbb{Z}_\ell} \mathbb{Q}_\ell$ carries a (continuous) representation of $\mathrm{Gal}(\overline{\mathbb{Q}}/\mathbb{Q})$, which is in fact a dual to $H^1(E(\overline{\mathbb{Q}}), \mathbb{Q}_\ell)$. That representation always has infinite image; indeed if you compose it with the determinant $\mathrm{Aut}_{\mathbb{Q}_\ell} V_\ell(E)) \to \mathbb{Q}_\ell^\times$, you find the ℓ-adic cyclotomic character $\chi_\ell : \mathrm{Gal}(\overline{\mathbb{Q}}/\mathbb{Q}) \to \mathbb{Z}_\ell^\times$, which is such that $\tau(\zeta) = \varepsilon^{\chi_\ell(\tau) \bmod \ell^n}$ for any root of unity ζ of order ℓ^n and any $\tau \in \mathrm{Gal}(\overline{\mathbb{Q}}/\mathbb{Q})$; the cyclotomic character has infinite image $\mathbb{Z}_\ell^\times$.

If moreover the elliptic curve E over $\mathbb{Q}$ has complex multiplication by an imaginary quadratic field I, then it is known that there exists a quasicharacter Θ of C_I, which is **not of finite order**, such that for any prime number ℓ the representation on $V_\ell(E)$ can be obtained from Θ (by an induction process, see § 3 on Weil groups) . (For elliptic curves the reader can consult [Lg 2, Hu, Kn, Si]).

All this shows not only that we should consider ℓ-adic representations besides complex representations, but also that some such representations could be connected with quasi-characters of idele class groups which are more general than finite order characters.

B.2.3 The continuous complex representations of $\mathrm{Gal}(\overline{F}/F)$, or its continuous ℓ-adic representations coming from geometry as above, share common properties.

In particular, if σ is such a representation on a space V, then for each place v of F, we can restrict σ to the subgroup $\mathrm{Gal}(\overline{F}_v/F_v)$ of $\mathrm{Gal}(\overline{F}/F)$, thus yielding a representation σ_v on the same space V, called the **component of σ ar v**. As the embedding of $\mathrm{Gal}(\overline{F}_v/F_v)$ into $\mathrm{Gal}(\overline{F}/F)$ is well-defined only up to inner automorphisms, only the **isomorphism class** of σ_v is well-defined – in fact we shall only really work with isomorphism classes of representations. Moreover, for almost all finite places v of F, σ_v is **unramified**, i.e. trivial on the inertia group I_{F_v}, in which case σ_v is determined by a unique automorphism $\sigma_v(\Phi_{F_v})$ of V – again only its conjugacy class in $\mathrm{Aut}(V)$ is well-defined. In fact it is known that if σ is semisimple then the conjugacy classes of $\sigma_v(\Phi_{F_v})$'s for almost all v (or even their traces) determine σ up to isomorphism [Se, I.2.3].

B.2.4 What Langlands proposed already some time ago [Ll 1], and which has been precised and refined over the years, is to relate Galois representations as above to automorphic forms; more precisely n-dimensional representations of $\mathrm{Gal}(\overline{F}/F)$ should be related to automorphic representations of the locally compact group $\mathrm{GL}_n(\mathbb{A}_F)$ (see [Corv], the standard reference on automorphic forms and representations).

The topology on $\mathrm{GL}_n(\mathbb{A}_F)$ is defined as in the case of $\mathbb{A}_F$ or $\mathbb{A}_F^\times$; note that $\mathbb{A}_F^\times$ is none other that $\mathrm{GL}_1(\mathbb{A}_F)$. For a finite set S of finite places of F, containing all infinite places, we put the product topology on the subgroups $\mathrm{GL}_n(\mathbb{A}_F^S) = \Pi_{v\in S}\mathrm{GL}_n(F_v)\Pi_{v\notin S}\mathrm{GL}_n(\mathcal{O}_{F_v})$, and $\mathrm{GL}_n(\mathbb{A}_F)$ is the union of such subgroups with the inductive limit topology. For each place v of F, $\mathrm{GL}_n(F_v)$ embeds continuous into $\mathrm{GL}_n(\mathbb{A}_F)$, and we also have the diagonal embedding of $\mathrm{GL}_n(F)$ into $\mathrm{GL}_n(\mathbb{A})$, which makes $\mathrm{GL}_n(F)$ into a discrete subgroup of $\mathrm{GL}_n(\mathbb{A})$.

Automorphic forms for GL_n over F are functions on the homogeneous space $\mathrm{GL}_n(F)\backslash\mathrm{GL}_n(\mathbb{A}_F)$ which have some degree of smoothness and satisfy some growth conditions : see [BJ] for the exact requirements. The group $\mathrm{GL}_n(\mathbb{A}_F)$ acts naturally on fonctions on $\mathrm{GL}_n(F)\backslash\mathrm{GL}_n(\mathbb{A}_F)$, by right translations; strictly speaking however, it does not act on automorphic forms : the subgroup $\mathrm{GL}_n(\mathbb{A}_F)_f$ of elements with trivial component at infinite places does act, but at infinite places only the Lie algebra $\mathfrak{G}_\infty$ of the Lie group $\Pi_{v \text{ infinite}}\mathrm{GL}_n(F_v)$, and a fixed maximal compact subgroup K_∞, act.

B.2.5 Automorphic representations are (nearly) the irreducible representations of $\mathrm{GL}_n(\mathbb{A})$ occurring as subquotients of the representation on the space of automorphic forms; more precisely, since the group $\mathrm{GL}_n(\mathbb{A})$ does not act on automorphic forms, we have to consider only the joint action of $\mathrm{GL}_n(\mathbb{A}_F)_f$ and $\mathfrak{G}_\infty$, K_∞, and take an irreducible subquotient for that joint action.

Because $\mathrm{GL}_n(\mathbb{A}_F)$ is rather close to the product of the groups $\mathrm{GL}_n(F_v)$, an automorphic representation Π of $\mathrm{GL}_n(\mathbb{A}_F)$ can be decomposed [Fl] as a kind of generalized tensor product $\Pi = \bigotimes_v \prod_v$, where Π_∞ the tensor product of the Π_v's for infinite v is a $(\mathfrak{G}_\infty, K_\infty)$-module, whereas, for a finite place v, Π_v is a **smooth irreducible** representation of $\mathrm{GL}_n(F_v)$.

Then Π_v is called the component of Π at v; it is well-defined only up to isomorphism.

Here a **smooth** representation of $\mathrm{GL}_n(F_v)$ on a complex vector space W is a homomorphism of $\mathrm{GL}_n(F_v)$ into $\mathrm{Aut}_{\mathbb{C}}(W)$ such that every vector w in W has open stabilizer in $\mathrm{GL}_n(F_v)$. (Of course, the same definition applies to representations of $\mathrm{GL}_n(K)$ for our p-adic field K). This smoothness condition is very strong and makes possible a nearly completely algebraic investigation of smooth representations see, e.g [BZ, Ct, Z]. The smooth representation is irreducible if W is non zero and has no subspace, other than W and $\{0\}$, invariant under $\mathrm{GL}_n(F_v)$.

If $\Pi = \bigotimes_v \prod_v$ is an automorphic representation of $\mathrm{GL}_n(\mathbb{A}_F)$ as above, then for almost all finite places v of F, Π_v is **unramified**, meaning that there is a non-zero vector in the space of Π_v which is fixed under the maximal compact subgroup $\mathrm{GL}_n(\mathcal{O}_{F_v})$ of $\mathrm{GL}_n(F_v)$. Such smooth irreducible unramified representations of $\mathrm{GL}_n(F_v)$ have been classified by Satake (see [Ct]) : their isomorphism classes are parametrized by n-tuples $(\alpha_1, \dots, \alpha_n)$ of complex numbers, up to ordering.

Note : most automorphic representations of $\mathrm{GL}_n(\mathbb{A}_F)$, and most smooth irreducible representations of $\mathrm{GL}_n(F_v)$, are infinite dimensional when $n > 1$. When $n = 1$ however they all have dimension 1, and, for example, the isomorphism classes of automorphic representations of $\mathrm{GL}_1(\mathbb{A}_F)$ can be identified with the quasicharacters of $\mathbb{A}_F^\times$ trivial on $F^\times$; the component at the place v is simply the restriction to $F_v^\times$.

B.2.6 Semisimple representations of $\mathrm{Gal}(\overline{F}/F)$ are direct sums of irreducibles. In an analogous, but more complicated manner, all automorphic representations of $\mathrm{GL}_n(\mathbb{A}_F)$ can be obtained from building blocks, the **cuspidal** automorphic representations of $\mathrm{GL}_r(\mathbb{A}_F)$ for $r \leq n$: see [Ll 3]. A cuspidal automorphic representation Π of $\mathrm{GL}_n(\mathbb{A}_F)$ is characterized, up to isomorphic, by giving Π_v at almost all finite place v of F [JS 1].

Example : Let N be a positive integer. A classical modular form f of some weight $k \geq 1$ and level N [Lg 3, Kn] can be translated into an automorphic form on $\mathrm{GL}_2(\mathbb{A}_{\mathbb{Q}})$ [Ge]; if f is an eigenform for the Hecke operators T_p, $p \nmid N$, then the corresponding automorphic form gives rise to an automorphic representation $\Pi = \Pi_\infty \otimes \bigotimes_p \prod_p$ of $\mathrm{GL}_2(\mathbb{A}_{\mathbb{Q}})$; the component Π_∞ is determined by the weight k, whereas Π_p is unramified at primes $p \nmid N$, the parameters for Π_p being determined by the eigenvalue of T_p on f. The automorphic representation Π is cuspidal if and only if the modular form f is parabolic.

B.2.7 We can now express the conjectures of Langlands for GL_n over F, at least for complex Galois representations.

Conjecture 1. Let Σ be an irreducible, n-dimensional, complex representation of $\mathrm{Gal}(\overline{F}/F)$. Then there should exist an automorphic cuspidal $\Pi = \Pi(\Sigma)$ of $\mathrm{GL}_n(\mathbb{A}_F)$ **associated** to Σ in the sense that, for almost all finite places v of F, where both Σ_v and Π_v are unramified, the n-tuple of complex numbers $(\alpha_1(v), \dots, \alpha_n(v))$ parametrizing Π_v is the n-tuple of eigenvalues of $\Sigma_v(\Phi_{F_v})$.

Note that Π, if it exists at all, is determined up to isomorphism, and that two irreducible complex representations Σ and Σ' yielding the same automorphic cuspidal representation $\Pi(\Sigma) = \Pi(\Sigma')$ are isomorphic.

B.2.8 Of course, when $n = 1$, the conjecture is true because of class field theory. There are relatively few cases, when $n > 1$, where Conjecture 1 is known.

Most of those cases are for $n = 2$. When n is 2, the image of Σ in the projective linear group can be dihedral, or isomorphic to one of $\mathfrak{a}_4$, $\mathfrak{S}_4$, $\mathfrak{a}_5$. When that image is dihedral, the conjecture was proved long ago [JL]; when it is $\mathfrak{a}_4$, the conjecture was proved by Langlands [Ll 4] and for $\mathfrak{S}_4$ by Langlands and Tunnell [Tu]. Only recently has there been progress in the case where the projective image is $\mathfrak{a}_5$. Indeed recent work of Buzzard, Dickinson, Shepherd-Barron and Taylor gives the following result, when $F = \mathbb{Q}$.

THEOREM [BDST]. — *Let* $\Sigma : \mathrm{Gal}(\overline{\mathbb{Q}}/\mathbb{Q}) \to \mathrm{GL}_2(\mathbb{C})$ *be a continuous representation with projective image* $\mathfrak{a}_5$. *Assume that* Σ *is odd, i.e. that* $\det \Sigma$ *takes the value* -1 *on complex conjugations in* $\mathrm{Gal}(\overline{\mathbb{Q}}/\mathbb{Q})$. *Assume further either of* a) *or* b)

a) Σ_3 *and* Σ_5 *are unramified and* $\Sigma_5(\Phi_{\mathbb{Q}_5})$ *has order* 2 *in* $\mathrm{PGL}_2(\mathbb{C})$.

b) Σ_2 *and* Σ_3 *are unramified and* $\Sigma_2(\Phi_{\mathbb{Q}_2})$ *has order* 3 *in* $\mathrm{PGL}_2(\mathbb{C})$.

Then there is an automorphic Π *of* $\mathrm{GL}_2(\mathbb{A}_\mathbb{Q})$ *associated to* Σ, *which is in fact attached to a parabolic modular form of weight* 1 *as above.*

The only general case where $\Pi(\Sigma)$ is known to exist, for $n > 2$, is when the image of Σ is nilpotent [AC, III].

B.2.9 It is clear from class field theory that one should expect a local counterpart to Conjecture 1. Let us first consider finite places; we then get the following conjecture for the p-adic field K.

Conjecture 2 . There exists a **natural** way to attach, to each isomorphism class σ of n-dimensional complex representations of $\mathrm{Gal}(\overline{K}/K)$, an isomorphism class $\pi = \pi(\sigma)$ of smooth irreducible representations of $\mathrm{GL}_n(K)$.

Note that we have not imposed any irreducibility condition on σ : indeed if Σ is an irreducible complex representation of $\mathrm{Gal}(\overline{F}/F)$, of dimension > 1, then Σ_v will in general be reducible, for example when it is unramified. In fact σ should be irreducible if and only if $\pi = \pi(\sigma)$ is supercuspidal i.e. has coefficients compactly supported mod. centre. As for number fields the supercuspidal representations are the building blocks which the construction of all smooth irreducible ones [BZ, Ct, Z].

That conjecture 2 is rather vague : we have to give a meaning to "natural". There are several ways to do that. Firstly, if σ is unramified, then certainly $\pi(\sigma)$ should be the class of smooth irreducible unramified representations of $\mathrm{GL}_n(K)$ with parameters given by the eigenvalues of $\sigma(\Phi_K)$. Seccondly, we want an even greater compatibility with conjecture 1 : if K is isomorphic to the completion F_v of F at some finite place v and Σ, $\Pi = \Pi(\Sigma)$ are as in conjecture 1, then the class of representations of $\mathrm{GL}_n(K)$ obtained from $\Pi(\Sigma)_v$ by transport of struccture should be $\pi(\sigma)$ where σ is the class of representations of $\mathrm{Gal}(\overline{K}/K)$ obtained from Σ_v.

B.2.10 In fact this suggests a way to prove conjecture 2, by a local-global principle : first prove conjecture 1 for as many Σ's as we can; then try and prove that the assignment of $\Pi(\Sigma)_v$ to Σ_v at finite places is purely local i.e. depends only on F_v and Σ_v and not on F and Σ. Then show that this allows to produce the natural map expected in conjecture 2.

This is indeed what works in [Ha 2, Ha T, He], but using ℓ-adic representations Σ instead of complex ones; indeed as explained above the supply of complex representations Σ for which $\Pi(\Sigma)$ is known to exist is too small; even, the existence of $\Pi(\Sigma)$, as in the theorem B.2.8, is often proved by a detour via ℓ-adic representations!

We shall come to ℓ-adic representations shortly, but we have to remark that it would be better, in conjecture 2, to express the naturality condition in purely local terms, rather than by the implied compatibility with any potential case of conjecture 1 (of course it would be even nicer to have a purely local proof of conjecture 2, but that is still a long way, I guess). Harris and Taylor [Ha T] in fact show that the cohomology of varieties constructed by Drinfeld provide a geometric model where one can see the correspondence $\sigma \mapsto \pi(\sigma)$ of conjecture 2, at least when σ is irreducible (see Part A, II)). They also show following [Ha 2], that the correspondence can be characterized in non-geometrical terms, by the preservation of some local invariants, the L and ε factors. Our paper [He] offers a simpler approach to that last characterization.

B.2.11 Before we turn to ℓ-adic representations, we should say what happens at the infinite places of F. At such an infinite place v, the local component of an automorphic representation of $\mathrm{GL}_n(\mathbb{A}_F)$ is not exactly a representation of $\mathrm{GL}_n(F_v)$; it is rather what is called an admissible irreducible $(\mathfrak{g}, K)$-module (see [Wa]). Langlands long ago classified [Ll 2] the admissible irreducible $(\mathfrak{g}, K)$-modules : for GL_n over the archimedean field F_v, their isomorphism classes are parametrized exactly by the isomorphism classes of n-dimensional representations of the **Weil group** W_{F_v}. The Weil group is bigger than the Galois group $\mathrm{Gal}(\overline{F}_v/F_v)$ (as we shall see later, there are versions over number fields and p-adic fields). When F_v is isomorphic to $\mathbb{C}$, the Weil group is simply $F_v^\times$; when F_v is isomorphic to $\mathbb{R}$, it is a non-trivial extension

$$1 \to \overline{F}_v^\times \to W_{F_v} \to \mathrm{Gal}(\overline{F}_v/F_v) \to 1$$

where $\mathrm{Gal}(\overline{F}_v/F_v)$ acts naturally on $\overline{F}_v^\times$.

Many irreducible admissible $(\mathfrak{g}, K)$-modules can occur as components at infinite places of automorphic cuspidal representations Π of $\mathrm{GL}_n(\mathbb{A}_F)$. However, if Π is associated to an irreducible complex representation Σ of $\mathrm{Gal}(\overline{F}/F)$ as in conjecture 1, then for each infinite place v of F, Π_v should correspond to Σ_v seen as a representation of W_{F_v} via the quotient map $W_{F_v} \to \mathrm{Gal}(\overline{F}_v/F_v)$, and indeed this is true whenever conjecture 1 has been established.

In any case, complex representations of $\mathrm{Gal}(\overline{F}/F)$ only account for those automorphic cuspidal representations with very special components af infinity.

B.2.12 Let us now turn to ℓ-adic representations of $\mathrm{Gal}(\overline{F}/F)$. For us here, an ℓ-adic representation of $\mathrm{Gal}(\overline{F}/F)$ will be a continuous homomorphism $\Sigma : \mathrm{Gal}(\overline{F}/F) \to \mathrm{Aut}_{\overline{\mathbb{Q}}_\ell}(V)$, where V is a finite dimensional $\overline{\mathbb{Q}}_\ell$ vector space, satisfying the following properties

(i) V is defined over a finite extension of $\mathbb{Q}_\ell$;

(ii) V is unramified at almost all finite places of F (see [Ra] for the necessity of that last hypothesis).

Certainly the representations coming from ℓ-adic étale cohomology of smooth and proper varieties over F satisfy those properties.

Isomorphism, irreducibility, semisimplicity have the usual meaning. A semisimple ℓ-adic representation Σ of $\mathrm{Gal}(\overline{F}/F)$ is determined by the data of the traces $\mathrm{tr}\,\Sigma_v(\Phi_{F_v})$, for almost all (unramified) places v of F.

One would like to relate such an ℓ-adic representation Σ, at least when irreducible, to an automorphic cuspidal representation Π of $\mathrm{GL}_n(\mathbb{A}_F)$ by relating, at unramified places, the eigenvalues of $\Sigma_v(\Phi_{F_v})$ to the parameters for Π_v. However, several difficulties arise :

(i) The eigenvalues of $\Sigma_v(\Phi_{F_v})$ are in $\overline{\mathbb{Q}}_\ell$ whereas the parameters for Π_v are complex numbers;

(ii) The irreducible ℓ-adic representations Σ with an associated Π should satsify some condition bearing on the restriction Σ_v at the places v of F **above ℓ.**

(iii) Accordingly the automorphic cuspidal representations Π associated to a Σ should satisfy some condition bearing on the components **at infinite places** of F.

B.2.13 Here I can give only brief indications. For details and precisions on (i) and (ii) see the work of Clozel [Cl 1] and for (iii) see, e.g., the report of Fontaine and Mazur [FM].

The theory of ℓ-adic representations of $\mathrm{Gal}(\overline{K}/K)$, when K is a p-adic field with $p \neq \ell$, is very close to the theory of complex representations of $\mathrm{Gal}(\overline{K}/K)$ [De 2, Ta 2]. The situation is entirely different when $p = \ell$, essentially because $\mathrm{Gal}(\overline{K}/K)$ has then a very large pro-ℓ-normal subgroup, the wild ramification group. There is a whole hierarchy of ℓ-adic representations for ℓ-adic fields; we refer to [Per-p] and [FM]. In any case, the condition in (ii) above is that at places v of F above ℓ, Σ_v should be a **potentially semistable** representation.

On the other hand, the condition in (iii) above is that, for any infinite place v of F, Π_v should be **algebraic** i.e. parametrized by an algebraic complex representation $\sigma(\Pi_v)$ of W_{F_v} ; here $\sigma(\Pi_v)$ is **algebraic** if its restriction to the subgroup $\overline{F}_v^\times \simeq \mathbb{C}^\times$ is a sum of algebraic characters of the form $z \mapsto z^a \bar{z}^b$ for $z \in \mathbb{C}^\times$ with integers a, b. In fact one expects a very close relationship between the components of Π at infinite places (e.g. the exponents a, b in the algebraic characters occurring in $\sigma(\Pi_v)$) and the components of Σ at places above ℓ (e.g. the Hodge-Tate numbers). I refer to [Cl 1] for such considerations. Actually this is not quite $\sigma(\Pi_v)$ which should be algebraic but a twist of it by some positive real valued character. I shall neglect that complication here, although it can be annoying sometimes cf. [Ha T § 11], [He SA § 5].

B.2.14 Suppose now that Π is an automorphic cuspidal representation of $\mathrm{GL}_n(\mathbb{A}_F)$ which is algebraic at infinite places in the above sense. Then it is expected that Π is defined over a number field; more precisely there should be a number field E in $\mathbb{C}$ such that the tensor product Π_f of the Π_v's, over finite places v, has a model over E. This is in fact proved by Clozel [Cl 1] when Π_v verifies some further regularity condition at infinity. Moreover, there should be, for any finite place λ of E of residue characteristic ℓ, say, a continuous representation $\Sigma_\lambda = \Sigma(\Pi)_\lambda : \mathrm{Gal}(\overline{F}/F) \to \mathrm{GL}_n(E_\lambda)$ which is associated to Π is the sense that at every finite place v of F, not above ℓ, where Π_v is unramified, $\Sigma_{\lambda,v}$ is also unramified, the characteristic polynomial of $\Sigma_{\lambda,v}(\Phi_{F_v})$ being in $E[T]$ and equal (in $\mathbb{C}[T]$) to $\Pi_{i=1}^n(1 - \alpha_i(v)T)$ where $(\alpha_1(v), \ldots, \alpha_n(v))$ in the n-tuple parametrizing Π_v. (Again, to be entirely correct, there is a slight twist which I shall neglect here). Of course one also expects that $\Sigma_{\lambda,v}$ is potentially semistable at finite places v above the residue characteristic ℓ of E_λ, and that there is some precise relationship between $\Sigma_{\lambda,v}$ and Π_v, Π_∞, at such a place v.

Conversely any ℓ-adic representation Σ as in B.2.12, which is moreover potentially semistable at places above ℓ, should be obtained from some $\Sigma(\Pi)_\lambda$ as above, by extending the scalars from E_λ to $\overline{\mathbb{Q}}_\ell$.

B.2.15 In fact it is presumed that the converse statement is much more difficult than the direct one. For example, when $n = 2$, it has been known for some time how to attach ℓ-adic representations to classical modular forms of weight $k \geq 2$ [De 1] or even $k = 1$ [DS]. But it is only recently, through the works of Wiles [Wi], Taylor-Wiles [TW] and Breuil-Conrad-Diamond-Taylor [BCDT], that the Shimura-Tamiyama-Weil conjecture was proved, saying that each elliptic curve over $\mathbb{Q}$ is **modular**. In our terms :

THEOREM. — *Let E be an elliptic curve over $\mathbb{Q}$. Then there exists a modular parabolic newform f of weight 2, with rational, even integral, coefficients, such that the cuspidal automorphic representation Π of $\mathrm{GL}_2(\mathbb{A}_\mathbb{Q})$ attached to f has for each prime number ℓ an associated ℓ-adic representation $\Sigma(\Pi)_\ell$ of $\mathrm{Gal}(\overline{\mathbb{Q}}/\mathbb{Q})$, which is simply given by the action on $V_\ell(E) \otimes_{\mathbb{Q}_\ell} \overline{\mathbb{Q}}_\ell$.*

In more concrete terms, if $f(x) = \Sigma_{n\geq 1} a_n q^n$ is the development of f in $q = \exp(2\pi i z)$, then for each prime number p not dividing the level of f, E has good reduction at p and has exactly $p + 1 - a_p$ points over $\mathbb{Z}/p\mathbb{Z}$.

Remark. When $n = 1$, the conjectures of B.2.14 are true and can be deduced, with some work, from class field theory. See [Se Chap. II and He 1] for those and more generally for abelian ℓ-adic representations.

B.2.16 The result of Harris and Taylor quoted in part A (I.7, Global correspondence), which generalizes some earlier resultz of Kottwitz and Clozel [Ko, Cl 2, 3, CL], is of the sort looked for in B.2.15. Indeed, if F is a CM-field, and Π is an automorphic cuspidal representation of $GL_n(\mathbb{A}_F)$ satisfying some conditions – admittedly rather stringent but not precluding many examples – then there are associated Galois representations $\Sigma(\Pi)_\lambda$. The only catch is that there is an integer $a \geq 1$ such that dim $\Sigma(\Pi)_\lambda$ is *an* instead of the expected dimension n, and accordingly the characteristic polynomial of $\Sigma(\Pi)_\lambda(\Phi_{F_v})$ for unramified $\Sigma(\Pi)_{\lambda,v}$, is the a-th power of what it should be. There is some hope of getting Σ_λ's of the right dimension, by an argument of Taylor cf. [Ha 1 § 4]. In any case, in the application to the local correspondence, one can get Σ_λ's with the right dimension n (see, e;g. [He JA § 5]). We shall turn to that application in B.4, but only after discussing Weil groups for number fields and p-adic fields, and also L and ε-factors, which are all essential ingredients to the proof of the Langlands conjectures for GL_n over p-adic fields.

B.3. L and ε factors

B.3.1 It is now appropriate to say more about L and ε factors. They are essential invariants for Galois or automorphic representations, and they also provide crucial ingredients in the proofs. Emil Artin, already in the twenties, defined L-functions for continuous complex representations of $\mathrm{Gal}(\overline{F}/F)$. Nowadays, that can be presented as follows (see [Ta 2], for example).

If σ is a continuous representation of $\mathrm{Gal}(\overline{K}/K)$ (where K is our p-adic field), on a finite dimensional complex vector space V, then on the subspace V^{I_K} of vectors fixed by I_K, we get an unramified representation σ_I. The L factor of σ is defined, as a function of a complex parameter s, by $L(\sigma, s) = \det(\mathrm{Id} - \sigma_I(\Phi_K)|k_K|^{-s})^{-1}$ (this is the determinant of an endomorphism of V^{I_K}). This L factor is a rational function in $|k_K|^{-s}$, with complex coefficients.

Similarly it is possible to define L factors for continuous complex representations of $\mathrm{Gal}(\mathbb{C}/\mathbb{R})$ or $\mathrm{Gal}(\mathbb{C}/\mathbb{C})$; such factors involve the gamma function : for example the factor for the trivial character of $\mathrm{Gal}(\mathbb{C}/\mathbb{R})$ is $(2\pi)^{-s/2}\Gamma(s/2)$.

If F is our number field, and Σ a continuous complex representation of $\mathrm{Gal}(\overline{F}/F)$, we can form the **Artin L function**, given by the Euler product

$$L(\Sigma, s) = \Pi_v L(\Sigma_v, s), \quad \text{product over all places } v \ .$$

As a function of a complex variable, it converges absolutely for $\mathrm{Re}(s) > 1$. When F is $\mathbb{Q}$ and Σ is the trivial character, the product over prime numbers is none other than Riemann's zeta function. More generally if Σ is a character of $\mathrm{Gal}(\overline{\mathbb{Q}}/\mathbb{Q})$, it factorizes through the Galois group of a cyclotomic extension, hence corresponds (cf. B.1.7) to a character χ of $(\mathbb{Z}/N\mathbb{Z})^\times$ for some positive integer N, and the product over prime numbers is simply Dirichlet's L function $\Pi_{p\nmid N}(1 - \chi(p)p^{-s})^{-1}$.

B.3.2 Now it is well known that Riemann's zeta function, as well as Dirichlet's L functions, extend to meromorphic functions of s, with a functional equation relating s to $1-s$. The same is true for Artin L functions : for a general Σ as above, $L(\Sigma, s)$ extends to a meromorphic function on the whole complex plane, and if $\Sigma^\vee$ denotes the contragredient representation to Σ (the natural representation of $\mathrm{Gal}(\overline{F}/F)$ on the space dual to that of Σ), there is a functional equation

$$L(\Sigma; s) = \varepsilon(\Sigma, s)L(\Sigma^\vee, 1 - s),$$

where $\varepsilon(\Sigma, s)$ is a monomial αN^{-s} for some complex number α and some positive integer N. The way such things are proved is the following :

1) If Σ is one-dimensional, it corresponds by class field theory to a quasicharacter $\Pi(\Sigma)$ of $F^\times\backslash\mathbb{A}_F^\times$. For such a quasicharacter, an L-function can be defined which is known to have those properties (see below B.3.3); the equality $L(\Sigma, s) = L(\Pi(\Sigma), s)$, actually the equality of the corresponding factors at all places, implies the above properties for Σ.

2) If E is a finite extension of F in $\overline{F}$, and Σ' is a continuous complex finite dimensional representation of $\mathrm{Gal}(\overline{F}/E)$, then we can induce Σ' to $\mathrm{Gal}(\overline{F}/F)$ to get a representation $\Sigma = \mathrm{Ind}\,\Sigma'$ of $\mathrm{Gal}(\overline{F}/F)$. By construction, $L(\mathrm{Ind}\,\Sigma', s) = L(\Sigma, s)$;

there is even an equality for the local factors, corresponding to the extensions E_w/F_v of local fields.

3) By a theorem of Brauer, each continuous complex representation Σ of $\mathrm{Gal}(\overline{F}/F)$ is a linear combination $\Sigma = \Sigma_i n_i \operatorname{Ind}(\Sigma_i)$, where n_i is an integer, F_i a finite extension of F in $\overline{F}$, Σ_i a finite order character of $\mathrm{Gal}(\overline{F}/E_i)$. Accordingly we have

$$L(\Sigma, s) = \Pi_i L(\Sigma_i, s)^{n_i} ,$$

and meromorphicity and the functional equation follow.

Artin **conjectured** that $L(\Sigma, s)$ is even entire when Σ is **irreducible** of dimension > 1. As we shall see it is essentially a consequence of conjecture 1 of B.2.7.

B.3.3 I have just mentioned that a quasicharacter χ of $F^\times \backslash A_F^\times$ has an L-function with the right properties. In his thesis [CF, Lg 1], Tate gave a proof of that, where the global functional equation appears as a consequence of local functional equations for the component χ_v at all places v. This was progressively extended to L-function for automorphic representations of GL_n over F [GJ, JPSS 1], and even for a pair π, π' of automorphic representations of GL_n, $\mathrm{GL}_{n'}$ respectively [JPSS 2, JS 1, JS 2] see also [Sh 1, 2].

The pattern is the same in all situations. The first step is the functional equation for a p-adic field K (there is also a similar theory for infinite places, a bit more complicated). Let us describe it briefly for a pair π, π' of smooth irreducible representations of $\mathrm{GL}_n(K)$, $\mathrm{GL}_{n'}(K)$ respectively (to get the case of a single π, omit π'; to get the case of a quasicharacter of $K^\times$, take $n = 1$).

B.3.4 A space of integrals is defined, called zeta integrals, which depends on π, π' and auxiliary data. They are rational functions in q^{-s}, where $q = |k_K|$, and they generate a fractional ideal of $\mathbb{C}[q^s, q^{-s}]$ containing the constants. Such a fractional ideal has a unique generator of the form $P(q^{-s})^{-1}$ where $P \in \mathbb{C}[T]$, $P(0) = 1$. One puts $L(\pi \times \pi', s) = P(q^{-s})^{-1}$.

For each zeta integral Z as above, there is a "dual" zeta integral $\widehat{Z}$, and there is a non-zero monomial in q^{-s}, $\varepsilon(\pi \times \pi', s, \psi)$, such that for all Z we have

$$\frac{\widehat{Z}}{L(\pi^\vee \times \pi'^\vee, 1-s)} = \varepsilon(\pi \times \pi', s, \psi) \frac{Z}{L(\pi \times \pi', s)} .$$

Here ψ is a non-trivial character of the additive group K, and the process $Z \mapsto \widehat{Z}$ is some kind of Fourier transform involving ψ. Also $\pi^\vee$ denotes the representation contragredient to π i.e. the smooth irreducible representation of $\mathrm{GL}_n(K)$ on the space of linear functionals on the space of π which are **smooth**, i.e. fixed under an open subgroup of $\mathrm{GL}_n(K)$ (this is the natural duality for smooth representations).

B.3.5 The global functional equation reads, for Π, Π' automorphic representations of $\mathrm{GL}_n(\mathbb{A}_F)$, $\mathrm{GL}_{n'}(\mathbb{A}_F)$ respectively,

$$L(\Pi \times \Pi', s) = \varepsilon(\Pi \times \Pi', s) L(\Pi^\vee \times \Pi'^\vee, 1-s).$$

It is obtained as a consequence of the local functional equation and some kind of Poisson summation formula. The monomial $\varepsilon(\Pi \times \Pi', s)$, as the L factor, factorises as a product over all places

$$\varepsilon(\Pi \times \Pi', s) = \Pi_v \varepsilon(\Pi_v \times \Pi'_v, \Psi_v),$$

for any choice of non-trivial additive character Ψ of $\mathbb{A}_F$ trivial on F, Ψ_v being its restriction to F_v. A very important fact is that such L functions are very often **entire**; for example $L(\Pi, s)$ is entire if Π is a cuspidal automorphic representation of $\mathrm{GL}_n(\mathbb{A}_F)$, $n \geq 2$.

B.3.6 If χ is a quasicharacter of $\mathrm{Gal}(\overline{K}/K)$ corresponding to the quasicharacter $\pi(\chi)$ of $K^\times$ by class field theory, then we have $L(\chi, s) = L(\pi(\chi), s)$. Similarly over the number field F if χ is a quasicharacter of $\mathrm{Gal}(\overline{F}/F)$ corresponding to the quasicharacter $\pi(\chi)$ of $F^\times \backslash \mathbb{A}_F^\times$, then $L(\chi, s) = L(\pi(\chi), s)$ because of the equality of the factors at each place v. This gives step 1 in B.3.2.

In the general setting, if π is a smooth irreducible **unramified** representation of $\mathrm{GL}_n(K)$ with parameters $(\alpha_1, \dots, \alpha_n)$ then $L(\pi, s) = \Pi_{i=1}^n (1 - \alpha_i |k_K|^{-s})^{-1}$. If π' is another smooth irreducible **unramified** representation of $\mathrm{GL}_{n'}(K)$ with parameters $(\alpha'_1, \dots, \alpha'_n)$ then $L(\pi \times \pi', s) = \Pi_{i,1}^n (1 - \alpha_i \alpha'_j |k_K|^{-s})^{-1}$. Consequently, if σ (resp. σ') is an unramified complex representation of $\mathrm{Gal}(\overline{K}/K)$ such that the eigenvalues of Φ_K are $(\alpha_1, \dots, \alpha_n)$ (resp. $(\alpha'_1, \dots, \alpha'_n)$) then we have

$$L(\sigma, s) = L(\pi, s) \ , \ L(\sigma \otimes \sigma', s) = L(\pi \times \pi', s).$$

Therefore conjecture 1 in B.2.7 can be restated by saying that if Σ is an irreducible continuous complex representation of $\mathrm{Gal}(\overline{F}/F)$ of dimension n, there should exist a cuspidal automorphic representation $\Pi = \Pi(\Sigma)$ of $\mathrm{GL}_n(\mathbb{A}_F)$ with

$$L(\Sigma_v, s) = L(\Pi_v, s)$$

at almost all finite places v.

Of course one should expect that this holds at all places. In particular $L(\Sigma_v, s) = L(\Pi_v, s)$ should then be **entire** when $n \geq 2$, thus yielding **Artin's conjecture**, when one can associate $\Pi(\Sigma)$ to Σ. This is indeed the case for the 2-dimensional representations mentioned in B.2.8, for which Artin's conjecture is true.

B.3.7 In the context of conjecture 2 of B.2.9, it is to be expected that the correspondence $\sigma \mapsto \pi(\sigma)$ should satisfy $L(\pi(\sigma), s) = L(\sigma, s)$ and even

$$L(\pi(\sigma) \times \pi(\sigma'), s) = L(\sigma \otimes \sigma', s)$$

whenever σ, σ' are complex representations of $\mathrm{Gal}(\overline{K}/K)$.

Over the p-adic field K, the L factor are very often identically 1, hence do not give very informative invariants. However, from Tate's thesis, Langlands guessed that there are subtler invariants $\varepsilon(\sigma, s, \psi)$, depending on a non-trivial character ψ of K, and conjectured that they should be preserved under correspondence of conjecture 2

$$e(\pi(\sigma) \times \pi(\sigma'), s, \psi) = \varepsilon(\sigma \otimes \sigma', s, \psi).$$

Langlands' definition is very subtle.

Certainly, when σ is one dimensional, we can define $\varepsilon(\sigma, s, \psi)$ to be $\varepsilon(\pi(\sigma), s, \psi)$ where $\pi(\sigma)$ is the quasicharacter of $K^\times$ corresponding to σ. Also we ask

$$\varepsilon(\sigma \oplus \sigma', s, \psi) = \varepsilon(\sigma, s, \psi)\varepsilon(\sigma', s, \psi),$$

so that $\sigma \mapsto \varepsilon(\sigma, s, \psi)$ extends to a homomorphism from the Grothendieck group $R\mathcal{G}(K)$ of finite dimensional continuous complex representations of $\mathrm{Gal}(\overline{K}/K)$ to the multiplicative group of monomials in $|k_K|^{-s}$. Langlands then imposed the following further condition, inspired by step 2) in B.3.2, and called "inductivity in degree 0" : If L is a finite extension of K and $\rho \in \mathrm{R}\mathcal{G}(L)$ has degree 0, and if $\sigma = \mathrm{Ind}\ \rho$ is the element of $R\mathcal{G}(K)$ obtained by inducing ρ, then we should have

$$\varepsilon(\mathrm{Ind}\ \rho, s, \psi) = \varepsilon(\rho, s, \psi \circ \mathrm{tr}_{K/F}).$$

From Brauer's induction theorem already mentioned in B.3.2, it is clear that there is at most one way of defining such factors with the required properties. Langlands' existence proof was purely local, but was never published. Deligne found a much shorter proof, by a local-global principle [De, Ta 1]. This principle is based on the fact that if Σ is a complex representation of $\mathrm{Gal}(\overline{F}/F)$ then $\varepsilon(\Sigma, s)$ is the product of the local factors $\varepsilon(\Sigma_v, s, \Psi_v)$ where Ψ is as in B.3.5. A similar local-global principle lies behind the recent proofs of the local correspondence for GL_n i.e. of a precise version of conjecture 2 of B.2.9.

B.4. Weil groups, and the local correspondence

B.4.1 Before we give a precise – and proved – version of conjecture 2 of B.2.9, it is convenient to introduce the Weil group of the p-adic field K – Weil groups for number fields are used in the proofs, see below. For Weil groups and their representations, see [AT, § 15, De, Ta 2].

The Weil group W_K is a subgroup of $\mathrm{Gal}(\overline{K}/K)$ containing the inertia group I_K ; it is made out of the elements with image modulo I_K an integral power of the Frobenius Φ_K ; we give W_K the group topology for which I_K (with its topology as a Galois group) is open. The Weil group W_K is very close to $\mathrm{Gal}(\overline{K}/K)$ – it is dense in $\mathrm{Gal}(\overline{K}/K)$ – but it is slightly more flexible. Any representation of $\mathrm{Gal}(\overline{K}/K)$ gives a representation of W_K by restriction, but W_K has slightly more representations : in particular its quasicharacters are not necessarily of finite order, and by class field theory they are in bijection with **all** quasicharacters of $K^\times$.

Introducing the Weil group has the advantage of allowing us to state the Langlands conjectures for GL_n over K in terms of bijective maps, as we shall do now. To any continuous complex representation σ of W_K we attach an L factor $L(\sigma, s)$ and an ε factor $\varepsilon(\sigma, s, \psi)$, extending in the same manner the definition for representations of $\mathrm{Gal}(\overline{K}/K)$.

B.4.2 For each positive n, consider the set $\mathcal{G}_K^0(n)$ of isomorphism classes of irreducible continuous complex representations of W_K, of dimension n, and the set $\mathfrak{a}_K^0(n)$ of isomorphism classes of smooth irreducible supercuspidal representations of $\mathrm{GL}_n(K)$. (We have already mentioned in B.2.9 that, in a way similar to the global case, the supercuspidal representations are the building blocks for all smooth irreducible representations [BZ, Z]).

THEOREM [Ha T, He]. — *There is a unique family of bijections $\mathcal{G}_K^0(n) \to \mathfrak{a}_K^0(n)$, $\sigma \mapsto \pi(\sigma)$, satisfying the following properties :*

(i) *for $n = 1$, $\sigma \mapsto \pi(\sigma)$ is given by class field theory;*

(ii) *$\pi(\sigma^\vee) = \pi(\sigma)^\vee$ for $\sigma \in \mathcal{G}_K^0(n)$;*

(iii) *for* $\chi \in \mathcal{G}^0_K(1)$ *and* $\sigma \in \mathcal{G}^0_K(n)$, $\pi(\chi \otimes \sigma) = (\pi(\chi) \circ \det) \otimes \pi(\sigma)$;

(iv) *for* $\sigma \in \mathcal{G}^0_K(n)$ *and* $\sigma' \in \mathcal{G}^0_K(n')$, *then*

$$L(\pi(\sigma) \times \pi(\sigma'), s) = \lambda(\sigma \otimes \sigma', s)$$
$$\varepsilon(\pi(\sigma) \times \pi(\sigma'), s, \psi) = \varepsilon(\sigma \otimes \sigma', s, \psi)$$

for any non-trivial additive character ψ *of* F.

B.4.3 That theorem is proved by the local-global principle sketched in B.2.10, using the cases of the global correspondence – between ℓ-adic representations of $\mathrm{Gal}(\overline{F}/F)$ and automorphic cuspidal representations of $\mathrm{GL}_n(\mathbb{A}_F)$ – stated in A.1.7 and B.2.16. In the last part of this paper, I shall briefly sketch the argument, refering to [Ca] and [He JA] for more detail, and of course to the original papers [Ha 2, Ha T, He].

We start from the particular case of Theorem A.1.7 obtained from the analysis of the good reduction of Shimura varieties [Ko, Cl 2, Cl 3, CL] : the full theorem is a consequence of that special case and the local correspondence, see [Ha T § 11].

So we choose a totally real number field F_+ with some completion isomorphic to our p-adic field K; we also choose a quadratic imaginary field I split at p, and we consider the CM field $F = F_+ I$, which has a completion F_v isomorphic to K. By theorem A.1.7 we have a supply of cuspidal automorphic representations Π of $\mathrm{GL}_n(\mathbb{A}_F)$ which have **corresponding** ℓ-adic Galois representations $\Sigma_\ell : Gal(\overline{F}/F) \to \mathrm{GL}_{an}(\overline{\mathbb{Q}}_\ell)$, where a is a positive integer, hopefully equal to one. For each such cuspidal automorphic Π we would like to associate the class of the component Π_v to the class of the component $\Sigma_{\ell,v}$. The questions are as follows :

1) Is the assignment $\Pi_v \mapsto \Sigma_{\ell,v}$ purely local?

2) Does it preserve L and ε factors as in the theorem above?

3) Have we got enough global Π to get information on all of $\mathfrak{a}^0_K(n)$ and $\mathcal{G}^0_K(n)$?

Moreover there is the problem in questions 2) & 3) that Σ_ℓ is an ℓ-adic representation, and not a complex one!

B.4.4 Question 1 already is very serious. In [Ha T] it is proved, using geometry, that the assignment $\Pi_v \mapsto [\Sigma_{\lambda,v}]$, where $[\Sigma_{\lambda,v}]$ is the class of $\Sigma_{\lambda,v}$ in the Groghendieck group of finite dimensional ℓ-adic representations of W_{F_v}, is local. For a supercuspidal Π_v this is true because the correspondence $\Pi_v \mapsto [\Sigma_{\lambda,v}]$ can be seen in Drinfeld's geometric local model, which is local, and for general Π_v it comes out of the description of the bad reduction of Shimura varieties, which description form the essential part of [Ha T].

Another way of dealing with question 1 [He] is to show that if L and ε factors are preserved (question 2) and we have enough of the proposed local correspondance (question 3) then question 1 is also solved.

So we concentrate on questions 2 and 3, where Weil groups for number fields are used.

B.4.5 The only known way of dealing with question 2 is via the global functional equation for L functions. However, if L functions of automorphic representations – or pairs of such – are known to satisfy a functional equation (B.3.5), it is not known how to define L functions for ℓ-adic representations (especially the factors at places above ℓ and infinite places) so that they satisfy a functional equation. For ℓ-adic representations with finite image, one can use the complex L-functions, because a finite group has the "same" representation theory over all algebraically closed fields of characteristic zero, but ℓ-adic representations with finite image never appear in theorem A.I.7! So we need a larger supply of L functions with functional equations. This is provided by the (complex) representations of the Weil groups, to which we turn now.

B.4.6 A **Weil group** W_F for F is obtained as follows [AT § 15, Ta 2]. For any finite Galois extension E of F in $\overline{F}$ we have a relative Weil group $W(E/F)$, given by an extension of topological groups

$$1 \to C_E \to W(E/F) \to \mathrm{Gal}(E/F) \to 1\ ,$$

with the natural action of $\mathrm{Gal}(E/F)$ on the idele class group C_E, and W_F appears as a projective limit of the $W(E/F)$'s as E varies. An essential point here is that W_F mixes the Galois groups and the ideal class groups. In particular the quasicharacters of W_F are, as in the local case, in one-to-one correspondence with the quasicharacters of C_F. Also $\mathrm{Gal}(\overline{F}/F)$ is a quotient of W_F so that representations of $\mathrm{Gal}(\overline{F}/F)$ give rise to representations of W_F.

For each place v of F, finite or infinite, there is an embedding of the local Weil group W_{F_v} into W_F, well-defined up to – essentially – inner automorphism. If Σ is a continuous complex (finite dimensional) representation of W_F, we can define its components Σ_v by restriction to W_{F_v}, and Σ_v is unramified for almost all finite places v. We also define the global L function $L(\Sigma, s)$ as the product of $L(\Sigma_v, s)$ over all v's and its converges for $Re(s) >> 0$.

Of course if Σ comes from a representation of $\mathrm{Gal}(\overline{F}/F)$ we recover the previous definition. Now it is easy to see that Σ factorizes through some relative Weil group $W(E/F)$, and, using the fact that C_E is an abelian subgroup of finite index, that $L(\Sigma, s)$ is a quotient of products of factors $L(\chi, s)$ where χ is a quasicharacter of $C_{F'}$, F' an extension of F in E. It follows that $L(\Sigma, s)$ has a global functional equation $L(\Sigma, s) = \varepsilon(\Sigma, s)L(\Sigma^\vee, 1-s)$, where $\varepsilon(\Sigma, s)$ is the product of local ε factors $\varepsilon(\Sigma_v, s, \Psi_v)$ as in B.3.5.

B.4.7 It is only natural to say that a continuous complex representation Σ of W_F, of dimension n, and a cuspidal automorphic representation Π of $\mathrm{GL}_n(\mathbb{A}_F)$ **are associated** (or **correspond**) if for almost all places v of F, where Σ_v and Π_v are unramified, we have $L(\Sigma_v, s) = L(\Pi_v, s)$.

A crucial fact here is that if (Σ, Π), (Σ', Π') are two such corresponding pairs then [He 2 § 4, He]

$$L(\Sigma_v \otimes \Sigma'_v, s) = L(\Pi_v \times \Pi'_v, s)$$

$$\varepsilon(\Sigma_v \otimes \Sigma'_v, s, \Psi_v) = \varepsilon(\Pi_v \times \Pi'_v, s, \Psi_v)$$

at **all** finite places v.

Another important fact is that if Σ is **algebraic** (if Σ factorizes through $W(E/F)$ as above this means that its restriction to C_E is a sum of quasicharacters with algebraic component at infinity cf. B.2.13) then there is a number field E and for each finite place λ of E, a continuous representation $\mathrm{Gal}(\overline{F}/F) \to \mathrm{GL}_n(E_\lambda)$

which has the same characteristic polynomial of Frobenius at almost all places [Se, Ch. 3].

B.4.8 M. Harris was the one to remark that Theorem A.I.7 provides many instances where a cuspidal automorphic representation Π of $GL_n(\mathbb{A}_F)$ has associated ℓ-adic representations which in fact come from a complex continuous representation Σ of W_F, of dimension n. It follows that Σ and Π are associated.

Moreover Harris showed [Ha 2 § 4] that we can choose such pairs (Σ, Π) so that the classes $[\Sigma_v]$ we obtain generate, in the Grothendieck group of complex representations of W_{F_v}, a subgroup containing all representations factoring through a fixed finite Galois extension E_w/F_v. It follows that the assignment $\Pi_v \to [\Sigma_v]$ preserves L factors and ε factors as in Theorem B.4.2. With the geometric results of [Ha T] this proves the correspondence (theorem B.5.2) see [Ha T § 11]. My papers [He] and [He JA] offer a variant which uses only Theorem A.I.7 and not the hard geometry of [Ha T].

References

[AC] J. Arthur and L. Clozel, *Simple algebras, base change and the advanced theory of the trace formula,* **Annals of Math Studies, 120** Princeton University Press (1989).

[AT] E. Artin and J. Tate, *Clars Field Theory,* Benjamin (1968).

[BZ] J. Bernstein and A. Zelevinsky, *Induced representations of reductive p-adic groups I,* **Ann. Scient. Éc. Norm Sup. Paris, 10** (1977), 941–972.

[BJ] A. Borel and H. Jacquet, *Automorphic forms and automorphic representations,* in **[Corv]** Part I, 187–202.

[BHK] C. Bushnell, G. Henniart et P. Kutzko *Correspondance de Langlands locale pour* GL_n *et conducteurs de paires,* **Ann. Scient. Éc. Norm. Sup. (4)** 31 (1998), 537-560.

[BCDT] K. Buzzard, M. Dickinson, N. Shepherd-Barron, and R. Taylor, *On icosahedral Artin representations,* preprint.

[Ca] H. Carayol, *Preuve de la conjecture de Langlands locale pour* GL_n *: travaux de Harris-Taylor et Henniart,* Sém. Bourbaki n° 857, March 1999, **Astérisque** n° 266, SMF, Paris (2000), 191–244.

[Ca1] H. Carayol, *Non-abelian Lubin-Tate theory,* **in** L. Clozel and J.S. Milne eds., "*Automorphic forms, Shimura varieties and L-functions* II", **Academic Press (1990).**

[Ct] P. Cartier, *Representations of p-adic groups : a survey,* in **[Corv], Part I,** 111–156.

[CF] J. Cassels and A. Fröhlich, *Algebraic Number Theory,* **Academic Press,** (1967).

[Cl1] L. Clozel, *Motifs et Formes Automorphes : Applications du Principe de Fonctorialité* in **Automorphic Forms, Shimura Varieties, and L-functions, Vol. 1** L. Clozel and J. S. Milne, eds. **Perspectives in Mathematics 10** (1990), 77–160.

[Cl2] L. Clozel, *Représentations galoisiennes attachées aux représentations automrophes autoduales de* $GL(n)$ **Pub. Math. I.H.É. S. 73**(1994), 97–145.

[Cl3] L. Clozel, *On the cohomology of Kottwitz's arithmétic vaieties,* **Duke Math. J. 72** (1993), 757–795.

[CL] L. Clozel and J.-P. Labesse, *Changement de base pour les repésentations cohomologiques de certains groupes unitaires,"* **Astérisque** n^o257,SMF Paris (1999). 119–133

[Corv] A Borel and W. Casselman, eds. **Automorphic Forms, Representations, and L-functions,** P.S.P.M. **33,** A.M.S. (1979).

[De1] P. Deligne, *Formes modulaires et représentations p-adiques,* **Sém. Bourbaki** n^o**355, Lecture Notes in Math. 179** (1971), 55–105.

[De2] P. Deligne, *Les constantes des équations functionnelles des functions L* in **Modular forms II, Lecture Notes in Math. 349** Springer-Verlag (1973), 501–595.

[DS] P. Deligne and J.P. Serre, *Formes modulaires de poids 1* **Ann Scient. É N. S. 7**(1976), 507–530.

[Fl] P. Flath, *Decomposition of representations into tensor products,* in **[Corl]**, Part I 179–184.

[FM] J.-M. Fontaine and B. Mazur, *Geometric Galois Representations,* in **Elliptic curves, modular forms and Fermat's last theorem,** International Press (1995), 41–78.

[Ge] S. Gelbart, *Automorphic forms on adele groups,* **Annals of Math Studies 83** Princeton University Press (1975).

[Ha1] M. Harris, *Supercuspidal representations in the cohomology of Drinfeld upper half spaces : elaboration of Carayol's program,* **Invent. Math 129** (1999), 75–120.

[Ha2] M. Harris, *The local Langlands conjecture for* $\mathrm{GL}_(n)$ *over a padic field,* $n < p$, **Invent. Math. 139** (1998), 177–210.

[HaT] M. Harris and R. Taylor, *On the geometry and cohomology of some simple Shimura varieties,* **Preprint** n^o227, Institut de Mathématiques de Jussieu, Paris, October, 1999.

[He1] G. Henniart, *Représentations l-adiques abéliennes,*in Séminaire de Théorie des Nombres de Paris 1980-81, M.-J. Bertin, ed. **Progress in Mathematics 22** (1982), 107–126.

[He2] G. Henniart, *On the local Langlands conjecture for* $\mathrm{GL}_(n)$ *: the cyclic case,* **Annals of Math 123** (1986), 145–203.

[He] G. Henniart, Une preuve simple de la conjecture de Langlands pour GL_n sur un corps *p*-adique, **Invent. Math. 139** (1999), 439–455.

[HeJA] G. Henniart, Sur la conjecture de Langlands locale pour GL_n, exposé aux Journées Arithmétiques de Rome 1999, **à paraître au Journal de Théorie des Nombres de Bordeaux.**

[Hu] D. Husemoller, *Elliptic curves,* **GTM111** Springer-Verlag (1986).

[JL] H. Jacquet and R.P. Lanlands, *Automorphic forms on* $\mathrm{GL}_(2)$, **Lecture Notes in Math 114** Springer-Verlag (1986).

[JPSS1] H. Jacquet, I. I. Piatetski-Shapiro, and J. Shalika, *Automrophic forms on* $\mathrm{GL}_(3)$, *I and II*, **Annals of Math 109** (1979), 169–258.

[JPSS2] H. Jacquet, I. I. Piatetski-Shapiro, and J. Shalika, *Rankin-Selberg convolutions,* **Amer. J. Math. 105**(1983), 367–469

[JS1] H. Jacquet and J. Shalika, *On Euler products and the classification of automorphic representations I,* **Amer. J. Math. 103** (1981), 449–558, and *II* **Amer. J. Math. 103** (1981), 777–815.

[JS2] H. Jacquet and J. Shalika, *Rankin-Selberg convolutions : Archimedian theory,* **Festschrift in Honor of I.I. Piatetski-Shapiro, part I,** Weizmann Science Press (1990), 125–207.

[Jz] Janucz, *Algebraic Number Fields,*Academic Press, (1973).

[Kn] A. Knapp, *Elliptic curves,* **Mathematical Notes, 40** Princeton University Press, (1992).

[Lg1] S. Lang*Algebraic Number Theory*, Addison Wesley, New York (1970).

[Lg2] S. Lang*Elliptic Functions*, Addison Wesley, New York (1973).

[Lg3] S. Lang*Introduction to Modular Forms*, Springer-Verlag, New York (1976).

[Ll1] R. P. Langlands, *Problems in the thoery of automorphic forms* in **Lectures in Modern Analysis and Applications, Lecture Notes in Math. 170,** Springer-Verlag (1970), 18–86.

[Ll2] R. P. Langlands, *On the classification of irreducible representations of real algebraic groups,* **IAS**, (1970), also in **Representation Theory and Harmonic Analysis on Semisimple Lie Groups, Math Surveys Monographs 31,** American Math. Soc. (1989), 101–170.

[Ll3] R. P. Langlands, *On the notion of an automorphic representation. A supplement to the preceding paper* **[B5]**, in **[Corv]**, Part I, 203–208.

[Ll4] R. P. Langlands, *Base change for* $\mathrm{GL}_{(2)}$, **Annals of Math. Studies** n^o96, Princeton University Press, (1980)

[MoT] Motives, U. Jannsen, S. Kleiman, J.-P. Serre, eds., **PSPM 55**, Part I and II.

[Ne] J. Neukirch, *Class field theory,* Springer-Verlag (1986).

[Per-p] *Périodes p-adiques* (Séminaire de Bures 1988) **Astèrisque** n^o233, SMF Paris (1994).

[Ra] R. Ramakrishna, *Infinitely ramified Galois representations,* **Ann Math 151** (2000), 793–815.

[Se] J.-P. Serre, *Abelian l-adic representations and elliptic curves,* Benjamin (1968).

[Sh1] F. Shahidi, *On certain L-functions,* **Amer. J. Math. 103** (1981), 297–355.

[Sh2] F. Shahidi, *Fourier transforms of intertwining operators and Plancherd measures for* $\mathrm{GL}_{(n)}$, **Amer. J. Math. 106** (1984), 67–111.

[Si] J. Silverman, *The Arithmetic of Elliptic curves,* Springer-Verlag (1986).

[Ta1] J. Tate, *Local constants,* in **Algebraic Number Fields,** A. Fröhlich ed., Academic Press, (1977), 89–131.

[Ta2] J. Tate, Number theoretic background, in **[Corv]**, Part II, 3–26.

[TW] R. Taylor and A. Wiles, *Ring-theoretic properties of certain Hecke algebras,* **Ann. Math. 191** (1995), 553–572.

[Tu] J. Tunnell, *Artin's conjeture for representations of octahedral type,* **Bull. A.M.S. 5** (1981), 173–175.

[We] A. Weil, *Basic Number Theory,* Springer-Verlag (1979).

[Wi] A. Wiles, *Modular elliptic curves and Fermat's last theorem,* **Ann. Math 191** (1995), 443–551.

[Z] A. Zelevinsky, *Induced representations of reductive p-adic groups II : On irreducible representations of* $\mathrm{GL}_{(n)}$ **Ann. Scient. É. N.S.** Paris **13** (1980), 165–210.

Département de Mathématiques de l'Université Paris-Sud, et UMR 8628 du CNRS, Bâtiment 425, 91405 ORSAY Cedex France,

email : Guy.Henniart@math.u-psud.fr

Current Developments in Mathematics, 1999

The Langlands Correspondence for Function Fields following Laurent Lafforgue

Gérard Laumon

In June 1999, Laurent Lafforgue proved the Langlands correspondence for GL_r over a function field. His proof follows the strategy introduced by V. Drinfeld, more than 25 years ago, in the rank 2 case. In this lecture, I explain Lafforgue's theorem. I also sketch some of his argmuments in the everywhere unramified case.

I would like to thank S.-T. Yau for his kind invitation to the CDM conference and L. Lafforgue for his help during the preparation of these notes.

1. The statement

Let X be a smooth, projective and geometrically connected curve over a finite field $\mathbb{F}_q$ with q elements and let F be its function field.

We denote by $\mathbb{A} = \prod'_x F_x$ the topological ring of adeles of F. Here x runs through the set of places of F, or equivalently the set $|X|$ of closed points of X, and F_x is the completion of F at x. For each x we denote by $\mathcal{O}_x = \{a_x \in F_x \mid x(a_x) \geq 0\}$ the ring of integers of the local field F_x and by $\deg(x)$ the degree of its residue field $\kappa(x)$ over $\mathbb{F}_q$. There is a degree map

$$\deg : \mathbb{A}^\times \to \mathbb{Z},\ a \mapsto \sum_{x\in|X|} \deg(x)x(a_x)$$

which vanishes on $F^\times$ and $\mathcal{O}^\times$. It is well known that, for each $a \in \mathbb{A}^\times$ whose degree is non zero, the quotient $F^\times\backslash\mathbb{A}^\times/\mathcal{O}^\times a^{\mathbb{Z}}$ is finite.

Let $r \geq 1$ be an integer.

We first consider the adelic group $\mathrm{GL}_r(\mathbb{A})$. As usual we identify its center (the subgroup of scalar matrices) with $\mathbb{A}^\times$. The space of *cuspidal automorphic forms*

$$L_{\mathrm{cusp}} = L_{\mathrm{cusp}}(\mathrm{GL}_r(F)\backslash \mathrm{GL}_r(\mathbb{A}))$$

is by definition the space of complex functions φ on $\mathrm{GL}_r(\mathbb{A})$ which satisfy the following properties:

- - $\varphi(\gamma g) = \varphi(g)$, $\forall\gamma \in \mathrm{GL}_r(F)$, $\forall g \in \mathrm{GL}_r(\mathbb{A})$,
- - there exists a subgroup $K_\varphi \subset K := \mathrm{GL}_r(\mathcal{O}) = \prod_x \mathrm{GL}_r(\mathcal{O}_x)$ of finite index such that $\varphi(gk) = \varphi(g)$, $\forall g \in \mathrm{GL}_r(\mathbb{A})$, $\forall k \in K_\varphi$,
- - there exists $a \in \mathbb{A}^\times$ such that $\deg(a) \neq 0$ and $\varphi(ga) = \varphi(g)$, $\forall g \in \mathrm{GL}_r(\mathbb{A})$,

- - for every non trivial partition $r = r_1 + \cdots + r_s$ defining a standard parabolic subgroup $P = MU \subsetneq \mathrm{GL}_r$ with unipotent radical U and Levi component $M \cong \mathrm{GL}_{r_1} \times \cdots \times \mathrm{GL}_{r_s}$, we have

$$\int_{U(F)\backslash U(\mathbb{A})} \varphi(ug)du = 0, \ \forall g \in \mathrm{GL}_r(\mathbb{A}),$$

where du is any Haar measure on $U(F)\backslash U(\mathbb{A})$.

The *Hecke algebra* $\mathcal{H} = C_c^\infty(\mathrm{GL}_r(\mathbb{A}))$ is the convolution algebra of locally constant functions with compact support on $\mathrm{GL}_r(\mathbb{A})$. It acts on L_{cusp} by right convolution.

The *cuspidal automorphic representations of* $\mathrm{GL}_r(\mathbb{A})$ are by definition the simple $\mathcal{H}$-modules which occur as subquotients of L_{cusp}. We will denote by $\mathcal{A}_r$ the set of (isomorphism classes of) these representations. Any $\pi \in \mathcal{A}_r$ admits a *central character* $\omega_\pi : F^\times\backslash\mathbb{A}^\times \to \mathbb{C}^\times$ which is of finite order as we have $\omega_\pi(a) = 1$ for some $a \in \mathbb{A}^\times$ of non zero degree.

The Hecke algebra is the restricted tensor product of local Hecke algebras $\mathcal{H}_x$. Accordingly any $\pi \in \mathcal{A}_r$ is a restricted tensor product of simple $\mathcal{H}_x$-modules π_x.

Let $e_{K_x} \in \mathcal{H}_x$ be the characteristic function of the standard maximal compact subgroup $K_x = \mathrm{GL}_r(\mathcal{O}_x) \subset \mathrm{GL}_r(F_x)$. For each $\pi \in \mathcal{A}_r$ the set N_π of *ramified places* of π is by definition the finite set of places x such that

$$\pi_x * e_{K_x} = (0).$$

For any $x \notin N_\pi$ the complex vector space $\pi_x * e_{K_x}$ is an irreducible module over the commutative algebra

$$e_{K_x} * \mathcal{H}_x * e_{K_x} \cong \mathbb{C}[z_1, z_1^{-1}, \ldots, z_r, z_r^{-1}]^{\mathfrak{S}_r}.$$

Therefore, it is one dimensional and its isomorphism class is completely determined by an unordered r-tuple

$$(z_{x,1}(\pi), \ldots, z_{x,r}(\pi))$$

of complex numbers, the so-called *Hecke eigenvalues* of π at x, or equivalently by the sequence of power sums

$$S_x^{(n)}(\pi) = z_{x,1}(\pi)^n + \cdots + z_{x,r}(\pi)^n, \ n \geq 1,$$

or by the *local L factor*

$$L_x(\pi, s) = \frac{1}{\prod_{i=1}^r (1 - z_{x,i}(\pi) q^{-s\deg(x)})}.$$

We fix a separable closure $\overline{F}$ of F and we denote by Γ_F the Galois group of $\overline{F}$ over F. We fix some prime number ℓ distinct from the characteristic of $\mathbb{F}_q$ and an algebraic closure $\overline{\mathbb{Q}}_\ell$ of $\mathbb{Q}_\ell$.

A *ℓ-adic representation* of Γ_F of rank r is a group homomorphism $\sigma : \Gamma_F \to \mathrm{GL}_r(\overline{\mathbb{Q}}_\ell)$ which has the following properties:

- - there exists $g \in \mathrm{GL}_r(\overline{\mathbb{Q}}_\ell)$ and a finite extension E_λ of $\mathbb{Q}_\ell$ in $\overline{\mathbb{Q}}_\ell$ such that $g\sigma(\Gamma_F)g^{-1} \subset \mathrm{GL}_r(E_\lambda) \subset \mathrm{GL}_r(\overline{\mathbb{Q}}_\ell)$,
- - $g\sigma g^{-1} : \Gamma_F \to \mathrm{GL}_r(E_\lambda)$ is continuous for the Krull topology on Γ_F and the ℓ-adic topology on $\mathrm{GL}_r(E_\lambda)$,

- - for all but finitely many $x \in |X|$, σ is *unramified at* x, i.e. the restriction σ_x of σ to any decomposition subgroup $D_x \subset \Gamma_F$ at x is trivial on the inertia subgroup $I_x \subset D_x$, and thus factors through the quotient $D_x/I_x \cong \Gamma_{\kappa(x)} = \mathrm{Frob}_x^{\widehat{\mathbb{Z}}}$, where Frob_x is the geometric Frobenius element in the Galois group $\Gamma_{\kappa(x)}$ of $\kappa(x)$.

We consider the set $\mathcal{G}_r$ of (isomorphism classes of) irreducible ℓ-adic representations σ of Γ_F of rank r, the determinant of which is of finite order. For each $\sigma \in \mathcal{G}_r$ we denote by N_σ its finite set of ramified places. For any $x \notin N_\sigma$ we denote by

$$(z_{x,1}(\sigma), \ldots, z_{x,r}(\sigma))$$

the unordered r-tuple of the eigenvalues of $\sigma_x(\mathrm{Frob}_x)$, by

$$S_x^{(n)}(\sigma) = z_{x,1}(\sigma)^n + \cdots + z_{x,r}(\sigma)^n, \ n \geq 1,$$

the corresponding sequence of power sums and by

$$L_x(\sigma, s) = \frac{1}{\prod_{i=1}^{r}(1 - z_{x,i}(\sigma)q^{-s\deg(x)})}$$

the corresponding *local L factor*.

We fix an isomorphism $\overline{\mathbb{Q}}_\ell \xrightarrow{\sim} \mathbb{C}$. Each time the ℓ-adic topology of $\overline{\mathbb{Q}}_\ell$ plays no role we will freely use this isomorphism to identify $\overline{\mathbb{Q}}_\ell$ with $\mathbb{C}$.

M AIN THEOREM (Langlands Correspondence). — *There exists a unique bijection*

$$\mathcal{A}_r \xrightarrow{\sim} \mathcal{G}_r, \ \pi \mapsto \sigma(\pi),$$

such that, for every $\pi \in \mathcal{A}_r$ *we have the equality of power sums*

$$S_x^{(n)}(\sigma(\pi)) = S_x^{(n)}(\pi), \ \forall n \geq 1,$$

or equivalently the equality of local L factors

$$L_x(\sigma(\pi), s) = L_x(\pi, s),$$

for all but finitely many places $x \notin N_{\sigma(\pi)} \cup N_\pi$.

For $r = 1$ this is a reformulation of the abelian class field theory in the function field case. For $r = 2$ the theorem has been proved by Drinfeld [3]. The general case $r \geq 3$ is due to Lafforgue [6].

For arbitrary r's but particular π's, some cases of the theorem had been proved earlier by Flicker and Kazhdan, and myself.

Remarks : (i) The uniqueness of the map $\mathcal{A}_r \xrightarrow{\sim} \mathcal{G}_r$ and its injectivity (assuming its existence) had been known for a long time. They respectively follow from the Čebotarev density theorem and from the strong multiplicity one theorem of Piatetski-Shapiro.

(ii) Let us fix an integer r. In order to prove the existence of the bijection $\mathcal{A}_{r'} \xrightarrow{\sim} \mathcal{G}_{r'}$ for $r' = 1, \ldots, r$ it is sufficient to prove the following weaker statement for $r' = 1, \ldots, r$:

$(\mathrm{A})_{r'}$ *For every* $\pi' \in \mathcal{A}_{r'}$ *there exists a Galois representation* $\sigma'(\pi') \in \mathcal{G}_{r'}$ *which satisfies the equality of local L factors*

$$L_x(\sigma'(\pi'), s) = L_x(\pi', s),$$

for all but finitely many places $x \notin N_{\pi'} \cup N_{\sigma'(\pi')}$.

Indeed, as was remarked by Deligne, if we have already proved the assertion $(\mathrm{A})_{r'}$ for $r' = 1, \dots, r-1$ the Grothendieck functional equation and the converse theorem of Hecke, Weil and Piatetski-Shapiro give for free the inverse maps

$$\mathcal{G}_{r'} \to \mathcal{A}_{r'},\ \sigma' \mapsto \pi'(\sigma'),$$

for $r' = 1, \dots, r$.

(iii) By standard techniques of L-functions one easily gets from the main theorem that $N_{\sigma(\pi)} = N_\pi$, that

$$L_x(\sigma(\pi), s) = L_x(\pi, s),$$

for all places $x \notin N_\pi$ and that, for each $x \in N_\pi$, the restriction of $\sigma(\pi)$ to any decomposition subgroup $D_x \subset \Gamma_F$ at x corresponds to π_x by the local Langlands correspondence. □

It is well known that the Jacquet-Shalika estimates of the Hecke eigenvalues of cuspidal automorphic representations and the main theorem imply the *Ramanujan-Petersson conjecture*:

T HEOREM (Drinfeld [2] for $r = 2$, Lafforgue [4], [6] for $r \geq 3$). — *For every $\pi \in \mathcal{A}_r$ and every place $x \notin N_\pi$ we have*

$$|z_{x,i}(\pi)| = 1,\ \forall i = 1, \dots, r.$$

□

There is a now standard strategy for constructing the map $\mathcal{A}_r \to \mathcal{G}_r$, $\pi \to \sigma(\pi)$.

The first step is to construct a "variety" V over F, equipped with an action of the Hecke algebra $\mathcal{H}$, so that its ℓ-adic cohomology

$$H_c^*(\overline{F} \otimes_F V, \overline{\mathbb{Q}}_\ell)$$

is a representation of the product of the Hecke algebra $\mathcal{H}$ and the Galois group Γ_F.

The second step is to compute the trace of this representation by the *Grothendieck-Lefschetz trace formula.*

The last step is to compare this geometric trace formula with the *Arthur-Selberg trace formula* in order to prove that the representation

$$\bigoplus_{\pi \in \mathcal{A}_r} \pi \otimes \sigma(\pi)$$

of $\mathcal{H} \times \Gamma_F$ that we are looking for occurs in $H_c^*(\overline{F} \otimes_F V, \overline{\mathbb{Q}}_\ell)$.

In the case we are considering there is an obstruction to the occurrence of the above direct sum representation into any ℓ-adic cohomology group. This strategy has thus to be slightly modified. Following Drinfeld it is the representation

$$\bigoplus_{\pi \in \mathcal{A}_r} \pi \otimes \sigma(\pi)^\vee \otimes \sigma(\pi)$$

of the product $\mathcal{H} \times \Gamma_F \times \Gamma_F$, where $\sigma^\vee$ is the contragredient representation of σ, which should occur in ℓ-adic cohomology.

Lafforgue proves the Langlands correspondence by induction on r. Assuming the Langlands correspondence $\mathcal{A}_{r'} \xrightarrow{\sim} \mathcal{G}_{r'}$ for all $1 \leq r' < r$ he constructs $\sigma(\pi)$ for each $\pi \in \mathcal{A}_r$.

Let $\mathcal{A}_r(K) \subset \mathcal{A}_r$ be the subset of everywhere unramified cuspidal automorphism representations π of $\mathrm{GL}_r(\mathbb{A})$ ($N_\pi = \emptyset$). *For simplicity we will only explain in this lecture Lafforgue's construction of* $\sigma(\pi)$ *for* $\pi \in \mathcal{A}_r(K)$.

2. Drinfeld shtukas

All the schemes (or stacks) that we will consider are over $\mathbb{F}_q$. *We simply denote by* $S \times T$ *the product over* $\mathbb{F}_q$ *of two schemes (or stacks). If* k *is a field which contains* $\mathbb{F}_q$ *and* S *is a scheme (or a stack), we will also use the notation* $k \otimes S = \mathrm{Spec}(k) \times S$. *For each scheme (or stack)* S *we denote by* Frob_S *its Frobenius endomorphism (relative to* $\mathbb{F}_q$*). For each scheme (or stack)* S *and each vector bundle* $\mathcal{E}$ *on* $S \times X$ *we define a new vector bundle* ${}^\tau\mathcal{E}$ *on* $S \times X$ *by*

$$ {}^\tau\mathcal{E} = (\mathrm{Frob}_S \times \mathrm{Id}_X)^*\mathcal{E}. $$

Let k be an algebraically closed field which contains $\mathbb{F}_q$.

A rank r vector bundle $\mathcal{E}$ on $k \otimes X$ equipped with an isomorphism ${}^\tau\mathcal{E} \xrightarrow{\sim} \mathcal{E}$ is nothing else than a rank r vector bundle on X. As it has been shown by Weil the set of isomorphism classes of rank r vector bundles on X is canonically isomorphic to the double coset space $\mathrm{GL}_r(F)\backslash \mathrm{GL}_r(\mathbb{A})/\mathrm{GL}_r(\mathcal{O})$.

D EFINITION (Drinfeld [1]). — *A (right)* shtuka $\widetilde{\mathcal{E}}$ *of rank* r *over* k *is a diagram*

$$ \mathcal{E} \xhookrightarrow{j_\infty} \mathcal{E}' \xhookleftarrow{j_o} {}^\tau\mathcal{E} $$

where:

- - $\mathcal{E}$ *and* $\mathcal{E}'$ *are two locally free* $\mathcal{O}_{k\otimes X}$*-Modules of rank* r*, or equivalently rank* r *vector bundles on* $k \otimes X$,
- - j_∞ *and* j_o *are two injective* $\mathcal{O}_{k\otimes X}$*-linear maps,*
- - *the torsion* $\mathcal{O}_{k\otimes X}$*-Modules* $\mathrm{Coker}(j_\infty)$ *and* $\mathrm{Coker}(j_o)$ *are of length* 1.

The supports $\infty, o \in X(k)$ *of* $\mathrm{Coker}(j_\infty)$ *and* $\mathrm{Coker}(j_o)$ *are called the* pole *and the* zero *of the shtuka.*

In other words a shtuka is a double modification of a rank r vector bundle $\mathcal{E}$,

$$ \mathcal{E} \xhookrightarrow{j_\infty} \mathcal{E}' \xhookleftarrow{j'_o} \mathcal{E}'' $$

(an elementary upper modification j_∞ at the point ∞ followed by an elementary lower modification j'_o at o), together with an isomorphism

$$ {}^\tau\mathcal{E} \xrightarrow{\sim} \mathcal{E}''. $$

The rank r shtukas are the points of a Deligne-Mumford algebraic stack Sht^r. The pole and the zero of the universal shtuka define a morphism

$$ (\infty, o) : \mathrm{Sht}^r \to X \times X $$

which is smooth of pure relative dimension $2r - 2$.

The stack Sht^r has infinitely many components $(\mathrm{Sht}^{r,d})_{d\in\mathbb{Z}}$ which are indexed by the degree of the universal shtuka

$$ \deg(\widetilde{\mathcal{E}}) = \deg(\mathcal{E}) = \deg(\mathcal{E}') - 1. $$

Example : For every integer d, the stack $\mathrm{Sht}^{1,d}$ is the fibered product

$$\begin{array}{ccc} \mathrm{Sht}^{1,d} & \longrightarrow & \mathrm{Bund}^{1,d} \\ \downarrow & \square & \downarrow L \\ X \times X & \xrightarrow[A]{} & \mathrm{Bund}^{1,0} \end{array}$$

where $\mathrm{Bund}^{1,d}$ is the Artin algebraic stack of line bundles of degree d on X, A is the Abel-Jacobi morphism which maps $(\infty, o) \in X(k) \times X(k)$ onto the line bundle $\mathcal{O}_{k\otimes X}(\infty - o)$, and L is the Lang "isogeny" which maps $\mathcal{L}$ onto $\mathcal{L}^{-1} \otimes_{\mathcal{O}_{k\otimes X}} {}^{\tau}\mathcal{L}$.

In particular, for every integer d the stack $\mathrm{Sht}^{1,d}$ is of finite type and admits a coarse moduli space which is a finite etale Galois covering of $X \times X$. □

But, except for $r = 1$ none of the components $\mathrm{Sht}^{r,d}$ *is of finite type.*

The Picard group $F^{\times}\backslash\mathbb{A}^{\times}/\mathcal{O}^{\times}$ of line bundles on X acts on the algebraic stack Sht^r: a line bundle $\mathcal{L}$ over X takes a rank r shtuka $\widetilde{\mathcal{E}}$ on $k \otimes X$ to

$$\mathcal{L} \otimes_{\mathcal{O}_X} \mathcal{E} \xhookrightarrow{\mathrm{Id}\otimes j} \mathcal{L} \otimes_{\mathcal{O}_X} \mathcal{E}' \xhookleftarrow{\mathrm{Id}\otimes t} \mathcal{L} \otimes_{\mathcal{O}_X} {}^{\tau}\mathcal{E} = {}^{\tau}(\mathcal{L} \otimes_{\mathcal{O}_X} \mathcal{E}).$$

This action does not change the pole and the zero.

Any $g \in \mathrm{GL}_r(\mathbb{A})$ defines a *Hecke correspondence*

$$c = (c_1, c_2) : \mathrm{Sht}^r(g) \to \mathrm{Sht}^r \times_{X\times X} \mathrm{Sht}^r,$$

where $\mathrm{Sht}^r(g)$ is a Deligne-Mumford algebraic stack and c_1, c_2 are etale representable morphisms. This correspondence only depends on the double coset $KgK \subset \mathrm{GL}_r(\mathbb{A})$ and does not change the pole and the zero. If N_g is the finite set of places x such that $g_x \notin F_x^{\times}K_x \subset \mathrm{GL}_r(F_x)$, c_1 and c_2 are finite over $((X \setminus N_g) \times (X \setminus N_g)) \times_{X\times X} \mathrm{Sht}^r$.

If $a \in \mathbb{A}^{\times}$ is a central element in $\mathrm{GL}_r(\mathbb{A})$ the corresponding Hecke operator is nothing else that the action of the element $F^{\times}a\mathcal{O}^{\times}$ of the Picard group of X.

3. Truncations

From now on we fix $a \in \mathbb{A}^{\times}$ such that $\deg(a) \neq 0$ *and we assume that* $r \geq 2$.

The quotient stack

$$\mathrm{Sht}^r/a^{\mathbb{Z}} \cong \coprod_{d=1}^{r\deg(a)} \mathrm{Sht}^{r,d}$$

has finitely many components, but is not of finite type. To study its ℓ-adic cohomology we will need to *truncate* it.

As for vector bundles on Riemann surfaces it is not difficult to define the *Harder-Narasimhan polygon* of a rank r shtuka $\widetilde{\mathcal{E}}$ over an algebraically closed field $k \supset \mathbb{F}_q$.

A *subobject* $\widetilde{\mathcal{F}}$ of $\widetilde{\mathcal{E}}$ is a pair of $\mathcal{O}_{k\otimes X}$-submodules $(\mathcal{F} \subset \mathcal{E}, \mathcal{F}' \subset \mathcal{E}')$ such that

- - $j(\mathcal{F}) \subset \mathcal{F}'$ and $t({}^{\tau}\mathcal{F}) \subset \mathcal{F}'$,
- - $\mathcal{E}/\mathcal{F}$ and $\mathcal{E}'/\mathcal{F}'$ are locally free $\mathcal{O}_{k\otimes X}$-modules of the same rank.

A subobject has a *rank*

$$\mathrm{rk}(\widetilde{\mathcal{F}}) = \mathrm{rk}(\mathcal{F}) = \mathrm{rk}(\mathcal{F}')$$

and, for each $\alpha \in \mathbb{R}$ an *α-degree*

$$\deg_\alpha(\widetilde{\mathcal{F}}) = (1-\alpha)\deg(\mathcal{F}) + \alpha\deg(\mathcal{F}').$$

If $\widetilde{\mathcal{F}}$ and $\widetilde{\mathcal{G}}$ are two subobjects of $\widetilde{\mathcal{E}}$ we say that $\widetilde{\mathcal{G}}$ is contained in $\widetilde{\mathcal{F}}$ and we write $\widetilde{\mathcal{G}} \subset \widetilde{\mathcal{F}}$ if $\mathcal{G} \subset \mathcal{F}$ and $\mathcal{G}' \subset \mathcal{F}'$.

Let $(0) = \widetilde{\mathcal{F}}_0 \subsetneq \widetilde{\mathcal{F}}_1 \subsetneq \cdots \subsetneq \widetilde{\mathcal{F}}_s = \widetilde{\mathcal{E}}$ be a filtration by subobjects of a shtuka. Its *α-polygon* is the continuous function

$$p : [0, r] \to \mathbb{R}$$

which vanishes at 0 and r, which is affine on the interval $[\mathrm{rk}(\widetilde{\mathcal{F}}_{j-1}), \mathrm{rk}(\widetilde{\mathcal{F}}_j)]$ for $j = 1, \ldots, s$, and which takes the values

$$p(\mathrm{rk}(\widetilde{\mathcal{F}}_j)) = \deg_\alpha(\widetilde{\mathcal{F}}_j) - \frac{\mathrm{rk}(\widetilde{\mathcal{F}}_j)}{r}\deg_\alpha(\widetilde{\mathcal{E}}),\ \forall j = 1, \ldots, s-1.$$

For a given $\alpha \in [0,1]$ and a given shtuka $\widetilde{\mathcal{E}}$ the set of the α-polygons of all the possible filtrations of $\widetilde{\mathcal{E}}$ admits a largest element $p_\alpha^{\mathrm{HN}}(\widetilde{\mathcal{E}})$, the so-called *Harder-Narasimhan polygon of index* α of the shtuka. It is a convex function.

The α's play a crucial role in Lafforgue's work. But in this lecture we will restrict ourself to $\alpha = 0$ and simply call *Harder-Narasimhan polygon* the polygon $p^{\mathrm{HN}} = p_0^{\mathrm{HN}}$.

We call *truncation parameter* any convex continuous function $p : [0, r] \to \mathbb{R}_{\geq 0}$ which vanishes at 0 and r and which is affine on each interval $[i-1, i]$ for $i = 1, \ldots, r$.

P ROPOSITION (Lafforgue [4]). — *For each truncation parameter p there exists a unique open substack*

$$\mathrm{Sht}^{r;\leq p} \subset \mathrm{Sht}^r$$

such that, for any algebraically closed field $k \supset \mathbb{F}_q$, we have

$$\mathrm{Sht}^{r;\leq p}(k) = \{\widetilde{\mathcal{E}} \in \mathrm{Sht}^r(k) \mid p^{\mathrm{HN}}(\widetilde{\mathcal{E}}) \leq p\}.$$

For every integer d and every truncation parameter p the open substack

$$\mathrm{Sht}^{r,d;\leq p} = \mathrm{Sht}^{r,d} \cap \mathrm{Sht}^{r;\leq p}$$

is of finite type. □

The open substacks $\mathrm{Sht}^{r;\leq p} \subset \mathrm{Sht}^r$ are obviously stable under the action of the Picard group of X. Therefore the algebraic stack $\mathrm{Sht}^r/a^{\mathbb{Z}}$ is an increasing union of open substacks of finite type

$$\mathrm{Sht}^{r;\leq p}/a^{\mathbb{Z}} \subset \mathrm{Sht}^r/a^{\mathbb{Z}}.$$

But none of the open substacks $\mathrm{Sht}^{r;\leq p}/a^{\mathbb{Z}}$ is stable under the action of the Hecke correspondences.

4. Lefschetz numbers

If ∞ and o are two closed points in X the finite subscheme $\infty \times o \subset X \times X$ has exactly $\delta(\infty, o)$ closed points, where $\delta(\infty, o)$ is the greatest common divisor of $\deg(\infty)$ and $\deg(o)$. For each $\xi \in \infty \times o$ the residue field $\kappa(\xi)$ is a composed extension of $\kappa(\infty)$ and $\kappa(o)$, and its degree $\deg(\xi)$ over $\mathbb{F}_q$ is thus the least common multiple

$$\mu(\infty, o) = \frac{\deg(\infty)\deg(o)}{\delta(\infty, o)}$$

of $\deg(\infty)$ and $\deg(o)$.

If ξ is a closed point in $X \times X$ we denote by $\mathrm{Sht}_\xi^r / a^{\mathbb{Z}}$ the fiber at ξ of the canonical projection $\mathrm{Sht}^r / a^{\mathbb{Z}} \to X \times X$. It is a smooth algebraic stack of pure dimension $2r-2$ over the finite field $\kappa(\xi)$. We denote by

$$\mathrm{Frob}_\xi : \mathrm{Sht}_\xi^r / a^{\mathbb{Z}} \to \mathrm{Sht}_\xi^r / a^{\mathbb{Z}}$$

its geometric Frobenius endomorphism relative to $\kappa(\xi)$.

Let us fix $g \in \mathrm{GL}_r(\mathbb{A})$. Let ∞ and o be two closed points in $X \setminus N_g$, let ξ be a closed point in $\infty \times o$, let n be a positive integer and let $p : [0, r] \to \mathbb{R}$ be a truncation parameter. We denote by

$$c_\xi = (c_{1,\xi}, c_{2,\xi}) : \mathrm{Sht}_\xi^r(g)/a^{\mathbb{Z}} \to \mathrm{Sht}_\xi^r / a^{\mathbb{Z}} \times_{\kappa(\xi)} \mathrm{Sht}_\xi^r / a^{\mathbb{Z}}$$

the fiber at ξ of the Hecke correspondence which is defined by g.

D EFINITION The Lefschetz number

$$\mathrm{Lef}(g \times \mathrm{Frob}_\xi^n, \mathrm{Sht}_\xi^{r;\leq p} / a^{\mathbb{Z}})$$

is the sum

$$\sum_y \frac{1}{|\mathrm{Aut}(y)|}$$

where y runs through the set of (isomorphism classes of) points in $\mathrm{Sht}_\xi^r(g)/a^{\mathbb{Z}}$ such that

$$c_{1,\xi}(y) = \mathrm{Frob}_\xi^n(c_{2,\xi}(y)) \in \mathrm{Sht}_\xi^{r;\leq p} / a^{\mathbb{Z}} \subset \mathrm{Sht}_\xi^r / a^{\mathbb{Z}},$$

and where $\mathrm{Aut}(y)$ is the finite automorphism group of the fixed point y.

We say that a truncation parameter is *convex enough* if, for every $i = 1, \ldots, r-1$ the slope of p on the interval $[i-1, i]$ is much bigger than the slope of p on the interval $[i, i+1]$.

Using Drinfeld's adelic description of shtukas, the particular case of the *fundamental lemma* proved by Drinfeld and the *Arthur-Selberg trace formula*, Lafforgue has shown:

. — *If* $\deg(\infty)$ *and* $\deg(o)$ *are large enough with respect to* g *and if* p *is convex enough with respect to* g*, the average Lefschetz number*

$$\frac{1}{\delta(\infty, o)} \sum_{\xi \in \infty \times o} \mathrm{Lef}(g \times \mathrm{Frob}_\xi^n, \mathrm{Sht}_\xi^{r;\leq p} / a^{\mathbb{Z}}).$$

is equal to the spectral expression

$$\sum_{\substack{\pi \in \mathcal{A}_r(K) \\ \omega_\pi(a)=1}} \mathrm{Tr}_\pi(f_g) q^{(r-1)n\mu(\infty,o)} S_\infty^{\left(-\frac{n\mu(\infty,o)}{\deg(\infty)}\right)}(\pi) S_o^{\left(\frac{n\mu(\infty,o)}{\deg(o)}\right)}(\pi)$$
$$+ \sum_{\substack{1 \leq r' < r \\ 1 \leq r'' < r}} \sum_{\substack{\pi' \in \mathcal{A}_{r'}(K) \\ \pi'' \in \mathcal{A}_{r''}(K)}} \mathrm{Tr}_{\pi',\pi''}^{\leq p}(f_g, n\mu(\infty, o)) S_\infty^{\left(-\frac{n\mu(\infty,o)}{\deg(\infty)}\right)}(\pi') S_o^{\left(\frac{n\mu(\infty,o)}{\deg(o)}\right)}(\pi'')$$

where

- - f_g the characteristic function of $KgK \subset \mathrm{GL}_r(\mathbb{A})$ and $\mathrm{Tr}_\pi(f_g)$ is the trace of the operator $\pi(f_g)$,
- - $m \mapsto \mathrm{Tr}^{\leq p}_{\pi',\pi''}(f_g, m)$ is a complex function of the integer m, which does not depend on the places $o, \infty \in X \setminus N_g$ and on the integer n, and which is of the form
$$\sum_{\lambda\in\Lambda} P_\lambda(m)\lambda^m$$
for some finite subset $\Lambda \subset \mathbb{C}^\times$ and some family $(P_\lambda(T))_{\lambda\in\Lambda}$ of polynomials in $\mathbb{C}[T]$,
- - the function $m \mapsto \mathrm{Tr}^{\leq p}_{\pi',\pi''}(f_g, m)$ is identically zero for all but finitely many pairs (π', π'').

(Recall that
$$S_x^{(m)}(\pi) = z_{x,1}(\pi)^m + \cdots + z_{x,r}(\pi)^m$$
is the m-th power sum of the Hecke eigenvalues of π at x.) □

If the open substacks $\mathrm{Sht}_\xi^{r;\leq p}/a^{\mathbb{Z}}$ of $\mathrm{Sht}_\xi^r/a^{\mathbb{Z}}$ were stable under the action of the Hecke operators, the main theorem would easily follows from the above result and would have been proved many years ago.

5. Compactifications

Let the torus $\mathbb{G}_{\mathrm{m}}^{r-1} = \mathrm{Spec}(\mathbb{F}_q[t_1, t_1^{-1} \dots, t_{r-1}, t_{r-1}^{-1}])$ act on the standard affine space $\mathbb{A}^{r-1} = \mathrm{Spec}(\mathbb{F}_q[u_1, \dots, u_{r-1}])$ by
$$(t_1, \dots, t_{r-1}) \cdot (u_1, \dots, u_{r-1}) = (t_1u_1, \dots, t_{r-1}u_{r-1}).$$
The quotient stack $[\mathbb{A}^{r-1}/\mathbb{G}_{\mathrm{m}}^{r-1}]$ is an Artin algebraic stack which is smooth of dimension 0. Its closed substack
$$\{u_1 \cdots u_{r-1} = 0\} = \{u_1 = 0\} \cup \cdots \cup \{u_{r-1} = 0\} \subset [\mathbb{A}^{r-1}/\mathbb{G}_{\mathrm{m}}^{r-1}]$$
is the union of $r - 1$ smooth divisors with normal crossings. The complementary open substack
$$\{u_1 \cdots u_{r-1} \neq 0\} = [\mathbb{G}_{\mathrm{m}}^{r-1}/\mathbb{G}_{\mathrm{m}}^{r-1}] \subset [\mathbb{A}^{r-1}/\mathbb{G}_{\mathrm{m}}^{r-1}]$$
is reduced to one point. For each partition $\mathbf{r} = (r_1 + \cdots + r_s = r)$ of r (into a sum of positive integers) the intersection
$$[\mathbb{A}_{\mathbf{r}}^{r-1}/\mathbb{G}_{\mathrm{m}}^{r-1}] = \bigcap_{i\in I}\{u_i = 0\} \cap \bigcap_{i\notin I}\{u_i \neq 0\}.$$
where we have set $I = \{r_1, r_1 + r_2, \dots, r_1 + \cdots + r_{s-1}\}$, is smooth of dimension $1 - s$. When $\mathbf{r}$ runs through the set of partitions of r the locally closed substacks $[\mathbb{A}_{\mathbf{r}}^{r-1}/\mathbb{G}_{\mathrm{m}}^{r-1}]$ form a stratification of $[\mathbb{A}^{r-1}/\mathbb{G}_{\mathrm{m}}^{r-1}]$.

Let us fix a truncation parameter p which is convex enough with respect to X. Let us also fix an integer d.

T HEOREM (Drinfeld [3] for $r = 2$, Lafforgue [5] for $r \geq 3$). — *There exists an Artin algebraic stack $\overline{\mathrm{Sht}}^{r,d;\leq p}$ and a stack morphism*
$$\overline{(\infty, o, \varepsilon)} : \overline{\mathrm{Sht}}^{r,d;\leq p} \to X \times X \times [\mathbb{A}^{r-1}/\mathbb{G}_{\mathrm{m}}^{r-1}]$$
with the following properties:

- - *all the automorphisms groups of* $\overline{\mathrm{Sht}}^{r,d;\leq p}$ *are* finite *(but not necessarily unramified),*
- - $\overline{(\infty,o,\varepsilon)}$ *is* smooth *of pure relative dimension* $2r-2$ *and*

$$\overline{(\infty,o)} = \mathrm{pr}_{X\times X}\circ\overline{(\infty,o,\varepsilon)} : \overline{\mathrm{Sht}}^{r,d;\leq p} \to X\times X$$

(which is also smooth of pure relative dimension $2r-2$*) is* proper,

- - *the restriction of* $\overline{(\infty,o,\varepsilon)}$ *over the open substack*

$$X\times X = X\times X\times[\mathbb{G}_{\mathrm{m}}^{r-1}/\mathbb{G}_{\mathrm{m}}^{r-1}] \subset X\times X\times[\mathbb{A}^{r-1}/\mathbb{G}_{\mathrm{m}}^{r-1}]$$

is nothing else than the stack morphism $(\infty,o):\mathrm{Sht}^{r,d;\leq p}\to X\times X.$ Λ

It follows from the theorem that the Artin algebraic stack $\overline{\mathrm{Sht}}^{r,d;\leq p}$ is proper and smooth of pure dimension $2r$, and that it contains the Deligne-Mumford algebraic stack $\mathrm{Sht}^{r,d;\leq p}$ as a dense open substack. The closed complementary substack is the union of $r-1$ divisors

$$\overline{\mathrm{Sht}}^{r,d;\leq p}\setminus\mathrm{Sht}^{r,d;\leq p} = \bigcup_{i=1}^{r-1}\overline{(\infty,o,\varepsilon)}^{-1}(\{u_i=0\}),$$

which are smooth with relative normal crossings over $X\times X$.

When $\mathbf{r}$ runs through the set of partitions of r the locally closed substacks

$$\overline{\mathrm{Sht}}_{\mathbf{r}}^{r,d;\leq p} = \overline{(\infty,o,\varepsilon)}^{-1}([\mathbb{A}_{\mathbf{r}}^{r-1}/\mathbb{G}_{\mathrm{m}}^{r-1}])$$

which are smooth of pure relative dimension $2r-1-s$ over $X\times X$, form a stratification of $\overline{\mathrm{Sht}}^{r,d;\leq p}$.

For each partition $\mathbf{r}=(r_1+\cdots+r_s=r)$ of r we also consider the Deligne-Mumford algebraic stack

$$\mathrm{Sht}^{\mathbf{r}} = \mathrm{Sht}^{r_1}\times_X\mathrm{Sht}^{r_2}\times_{X,\mathrm{Frob}_X}\mathrm{Sht}^{r_3}\times_{X,\mathrm{Frob}_X}\cdots\times_{X,\mathrm{Frob}_X}\mathrm{Sht}^{r_s}$$

which classifies the families $(\widetilde{\mathcal{E}}_1,\widetilde{\mathcal{E}}_2,\ldots,\widetilde{\mathcal{E}}_s)$ of shtukas of ranks $r_1,\ldots,r_s$ such that the zero o_1 of $\widetilde{\mathcal{E}}_1$ is equal to the pole ∞_2 of $\widetilde{\mathcal{E}}_2$ and that, for $j=2,\ldots,s-1$, the zero o_j of $\widetilde{\mathcal{E}}_j$ is equal to the image by the Frobenius endomorphism Frob_X of the pole ∞_{j+1} of $\widetilde{\mathcal{E}}_{j+1}$. By construction we have a smooth morphism of pure relative dimension $2r-2s$

$$(\infty_1,o_1=\infty_2,o_2=\mathrm{Frob}_X(\infty_3),\ldots,o_{s-1}=\mathrm{Frob}_X(\infty_s),o_s):\mathrm{Sht}^{\mathbf{r}}\to X\times X^{s-1}\times X.$$

Therefore $\mathrm{Sht}^{\mathbf{r}}$ is smooth of pure relative dimension $2r-s-1$ over $X\times X$.

For each $i=0,1,\ldots,r$ let $\widetilde{p}(i)$ be the unique integer in the length 1 interval

$$]p(i)+\frac{i}{r}d-1,p(i)+\frac{i}{r}d].$$

We set $d_1=\widetilde{p}(r_1)$ and we denote by $p_1:[0,r_1]\to\mathbb{R}$ the truncation parameter which takes the values

$$p_1(i_1)=\widetilde{p}(i_1)-\frac{i_1d_1}{r_1},\ \forall i_1=1,\ldots,r_1-1.$$

For each $j=2,\ldots,s$ we set

$$d_j=\widetilde{p}(r_1+\cdots+r_{j-1}+r_j)-\widetilde{p}(r_1+\cdots+r_{j-1})-1$$

and we denote by $p_j : [0, r_j] \to \mathbb{R}$ the truncation parameter which takes the values

$$p_j(i_j) = \widetilde{p}(r_1 + \cdots + r_{j-1} + i_j) - \widetilde{p}(r_1 + \cdots + r_{j-1}) - 1 - \frac{i_j d_j}{r_j}, \ \forall i_j = 1, \ldots, r_j - 1.$$

(The p_j's are essentially the normalized restrictions of p to the intervals $[r_1 + \cdots + r_{j-1}, r_1 + \cdots + r_{j-1} + r_j]$'s as in the figure below.

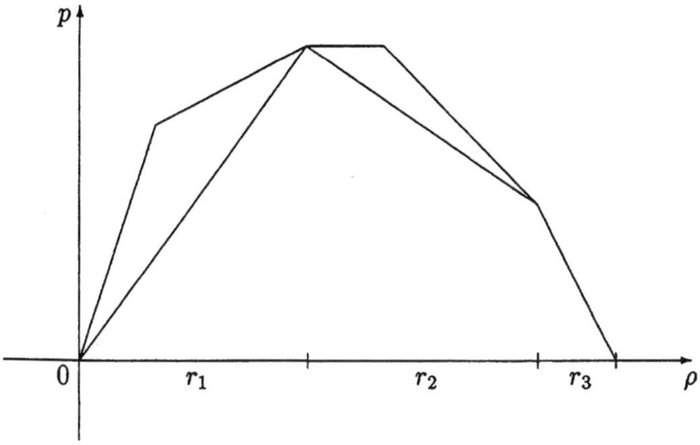

In particular all the p_j's are automatically convex enough with respect to X as soon as this is the case for p.)

We define an open substack

$$\mathrm{Sht}^{\mathbf{r},d;\,\leq p} \subset \mathrm{Sht}^{\mathbf{r}}$$

by requiring that, for $j = 1, \ldots, s$, the degree of the shtuka $\widetilde{\mathcal{E}}_j$ is equal to d_j and its Harder-Narasimhan polygon $p^{\mathrm{HN}}(\widetilde{\mathcal{E}}_j)$ is bounded above by p_j.

P ROPOSITION (Lafforgue [5]). — *For each non trivial partition* $\mathbf{r}$ *of* r *there exists a canonical morphism of stacks*

$$\overline{\mathrm{Sht}}_{\mathbf{r}}^{r,d;\,\leq p} \to \mathrm{Sht}^{\mathbf{r},d;\,\leq p}$$

which is the composition of a gerb whose structural group is finite, flat and radicial, and of a radicial representable morphism. Λ

Lafforgue calls *iterated shtukas* the points in $\overline{\mathrm{Sht}}^{r,d;\,\leq p}$. To each iterated shtuka is associated a partition $\mathbf{r} = (r_1, \ldots, r_s)$ of r and a family of "small" shtukas of ranks $r_1, \ldots, r_s$. Their zeros and poles $o_1 = \infty_2, o_2 = \mathrm{Frob}_X(\infty_3), \ldots, o_{s-1} = \mathrm{Frob}_X(\infty_s)$ are the *degenerators* of the iterated shtuka. The pole ∞_s and the zero o_1 are the *pole* and the *zero* of the iterated shtuka.

For each partition $\mathbf{r}$ of r we set

$$\overline{\mathrm{Sht}}_{\mathbf{r}}^{r;\,\leq p} = \coprod_{d \in \mathbb{Z}} \overline{\mathrm{Sht}}_{\mathbf{r}}^{r,d;\,\leq p} \subset \overline{\mathrm{Sht}}^{r;\,\leq p} = \coprod_{d \in \mathbb{Z}} \overline{\mathrm{Sht}}^{r,d;\,\leq p}$$

and

$$\mathrm{Sht}^{\mathbf{r};\leq p} = \coprod_{d\in\mathbf{Z}} \mathrm{Sht}^{\mathbf{r},d;\leq p}.$$

These algebraic stacks are naturally equipped with an action of the Picard group $F^\times\backslash\mathbb{A}^\times/\mathcal{O}^\times$ of X. In particular we may form the algebraic stack

$$\overline{\mathrm{Sht}}^{r;\leq p}/a^{\mathbf{Z}} = \coprod_{d=1}^{r\deg(a)} \overline{\mathrm{Sht}}^{r,d;\leq p}$$

which is a smooth compactification of $\mathrm{Sht}^{r;\leq p}/a^{\mathbf{Z}}$ over $X\times X$. It is stratified by the locally closed substacks $\overline{\mathrm{Sht}}_{\mathbf{r}}^{r;\leq p}/a^{\mathbf{Z}}$ which are "homeomorphic" to the Deligne-Mumford algebraic stacks $\mathrm{Sht}^{\mathbf{r};\leq p}/a^{\mathbf{Z}}$.

6. r-negligible Galois representations

In this section and the next one we will forget the action of the Hecke operators and concentrate on the Galois action on the ℓ-adic cohomology of the shtuka moduli varieties.

We denote by E the fraction field of $F\otimes F$ (the field of rational functions on the surface $X\times X$). We fix an algebraic closure $\overline{E}$ of E and an embedding $\overline{F}\otimes_{\overline{\mathbb{F}}_q}\overline{F}\hookrightarrow\overline{E}$ over the embedding $F\otimes F\hookrightarrow E$, where $\overline{\mathbb{F}}_q$ is the algebraic closure of $\mathbb{F}_q$ in $\overline{F}$.

We have thus defined a geometric point $\overline{\delta}:\mathrm{Spec}(\overline{E})\to X\times X$ over the generic point $\delta=\mathrm{Spec}(E)$ of $X\times X$. The images of $\overline{\delta}$ by the two canonical projections of $X\times X$ booth factors through the geometric point $\overline{\eta}:\mathrm{Spec}(\overline{F})\to X$ over the generic point η of X.

The Grothendieck fundamental group $\pi_1(X,\overline{\eta})$ is the quotient of $\Gamma_F = \mathrm{Gal}(\overline{F}/F)$ which classifies the finite extensions of F in $\overline{F}$ which are unramified everywhere. It admits as quotient the Galois group $\mathrm{Gal}(\overline{\mathbb{F}}_q/\mathbb{F}_q)$.

Similarly the Grothendieck fundamental group $\pi_1(X\times X,\overline{\delta})$ is a quotient of the Galois group $\mathrm{Gal}(\overline{E}/E)$ and admits $\mathrm{Gal}(\overline{\mathbb{F}}_q/\mathbb{F}_q)$ as quotient.

L EMMA The homomorphism

$$\pi_1(X\times X,\overline{\delta})\to\pi_1(X,\overline{\eta})\times\pi_1(X,\overline{\eta}),$$

which is induced by the two canonical projections of $X\times X$, is injective. Its image is the group of pairs of elements $\gamma',\gamma''\in\pi_1(X,\overline{\eta})$ which have the same images in the quotient $\mathrm{Gal}(\overline{\mathbb{F}}_q/\mathbb{F}_q)$. Λ

In particular, any irreducible ℓ-adic representation of $\pi_1(X\times X,\overline{\delta})$ is a direct factor of a semi-simple ℓ-adic representation of the form

$$\lambda\otimes(\sigma'\boxtimes\sigma'')$$

where λ is an ℓ-adic character of $\pi_1(X\times X,\overline{\delta})$ which factors through the quotient $\mathrm{Gal}(\overline{\mathbb{F}}_q/\mathbb{F}_q)\cong\widehat{\mathbb{Z}}$, and where σ' and σ'' are two irreducible ℓ-adic representations of $\pi_1(X,\overline{\eta})$ with determinants of finite order. We may identify λ to an ℓ-adic unit (its value at the geometric Frobenius element) and we may view the tensor product by λ as a generalized Tate twist.

We will simply call *virtual* $\pi_1(X \times X, \overline{\delta})$*-module* a formal linear combination $\sum_\rho m_\rho[\rho]$ where ρ runs through the set of (isomorphism classes of) irreducible ℓ-adic representations of $\pi_1(X \times X, \overline{\delta})$ and where the m_ρ's are rational numbers which are equal to 0 for all but finitely many ρ's. We say that ρ *occurs in* $\sum_\rho m_\rho[\rho]$ if its *multiplicity* m_ρ is non zero. The *trace* of $\sum_\rho m_\rho[\rho]$ is the function

$$\mathrm{Tr}_{\sum_\rho m_\rho[\rho]} : \pi_1(X \times X, \overline{\delta}) \to \overline{\mathbb{Q}}_\ell,\ \gamma \mapsto \sum_\rho m_\rho \,\mathrm{Tr}(\rho(\gamma)).$$

Any graded ℓ-adic representation H^* of $\pi_1(X \times X, \overline{\delta})$ defines a virtual $\pi_1(X \times X, \overline{\delta})$-module

$$[H^*] = \sum_\rho \sum_\nu (-1)^\nu m_\rho^\nu [\rho]$$

where m_ρ^ν is the number of times that ρ occurs in any Jordan-Hölder filtration of H^ν.

. — *A ℓ-adic representation of $\pi_1(X \times X, \overline{\delta})$ is said* r-negligible *if all its irreducible subquotients are direct factors of ℓ-adic representations of the form $\lambda \otimes (\sigma' \boxminus \sigma'')$ as above, where σ' and σ'' are both of dimension $\leq r-1$.*

A virtual $\pi_1(X \times X, \overline{\delta})$-module is said r-negligible *if any ρ which occurs in it is r-negligible.*

We now fix a truncation parameter which is convex enough with respect to X. We denote by $\mathrm{Sht}_{\overline{\delta}}^{r;\leq p}$, $\overline{\mathrm{Sht}}_{\overline{\delta}}^{r;\leq p}$ and $\mathrm{Sht}_{\overline{\delta}}^{\mathbf{r};\leq p}$ the fibers at the geometric point $\overline{\delta}$ of the morphisms (∞, o), $\overline{(\infty, o)}$ and (∞_1, o_s), and we consider their ℓ-adic cohomologies

$$H_c^*(r; \leq p) = H_c^*(\mathrm{Sht}_{\overline{\delta}}^{r;\leq p}/a^{\mathbb{Z}}, \overline{\mathbb{Q}}_\ell),$$

$$\overline{H}^*(r; \leq p) = H^*(\overline{\mathrm{Sht}}_{\overline{\delta}}^{r;\leq p}/a^{\mathbb{Z}}, \overline{\mathbb{Q}}_\ell),$$

$$H_\partial^*(r; \leq p) = H^*((\overline{\mathrm{Sht}}_{\overline{\delta}}^{r;\leq p} \setminus \mathrm{Sht}_{\overline{\delta}}^{r;\leq p})/a^{\mathbb{Z}}, \overline{\mathbb{Q}}_\ell)$$

and

$$H_c^*(\mathbf{r}; \leq p) = H_c^*(\mathrm{Sht}_{\overline{\delta}}^{\mathbf{r};\leq p}/a^{\mathbb{Z}}, \overline{\mathbb{Q}}_\ell).$$

It follows from the results of the previous section that:

- - the natural continuous actions of the Galois group $\mathrm{Gal}(\overline{E}/E)$ on these ℓ-adic cohomologies factor through the fundamental group $\pi_1(X \times X, \overline{\delta})$,
- - there is a long exact sequence

$$\cdots \to H_c^\nu(r; \leq p) \to \overline{H}^\nu(r; \leq p) \to H_\partial^\nu(r; \leq p) \to H_c^{\nu+1}(r; \leq p) \to \cdots$$

  and a spectral sequence

$$E_1^{s,\nu} = \bigoplus_{\mathbf{r}=(r_1,\ldots,r_{s+2})} H_c^\nu(\mathbf{r}; \leq p) \Rightarrow H_\partial^{\nu+s}(r; \leq p).$$

The virtual $\pi_1(X \times X, \overline{\delta})$-module

$$[H_c^*(r; \leq p)]$$

is thus equal to the virtual $\pi_1(X \times X, \overline{\delta})$-module

$$[\overline{H}^*(r; \leq p)] - \sum_{\mathbf{r}} [H_c^*(\mathbf{r}; \leq p)]$$

where $\mathbf{r}$ runs through the set of non trivial partitions of r.

7. The induction

Lafforgue proves the main theorem and the following proposition by a simultaneous induction on r.

P ROPOSITION (Lafforgue [6]). — *For any truncation parameter p which is convex enough with respect to X, the ℓ-adic representations $H_{\mathrm{c}}^{\nu}(r; \leq p)$ of $\pi_1(X \times X, \overline{\delta})$ is $(r+1)$-negligible for every integer ν.*

From now on we thus assume that the main theorem in rank $< r$ is already proved and that, for every truncation parameter p which is convex enough with respect to X, every positive integer $r' < r$ and every integer ν the $\pi_1(X \times X, \overline{\delta})$-module $H_{\mathrm{c}}^{\nu}(r'; \leq p)$ is r-negligible.

Let p be a truncation parameter which is convex enough with respect to X and let $\mathbf{r} = (r_1, \ldots, r_s)$ be a partition of r. Then it follows from the structure of the stratum $\mathrm{Sht}^{\mathbf{r}; \leq p}$ and the Künneth formula that, for each integer ν the ℓ-adic representation $H_{\mathrm{c}}^{\nu}(\mathbf{r}; \leq p)$ of $\pi_1(X \times X, \overline{\delta})$ is r-negligible. In fact the virtual $\pi_1(X \times X, \overline{\delta})$-module $[H_{\mathrm{c}}^{*}(\mathbf{r}; \leq p)]$ is a linear combination of virtual modules of the form

$$[\mu \otimes H_{\mathrm{c}}^{*}(\overline{\mathbb{F}}_q \otimes_{\mathbb{F}_q} X^{s-1}, \Theta_{j=1}^{s-1}(\sigma_j'' \otimes \sigma_{j+1}')) \otimes (\sigma_1' \boxminus \sigma_s'')] = \sum_{\lambda} m_{\lambda} [\lambda \otimes (\sigma_1' \boxminus \sigma_s'')]$$

where μ and λ are ℓ-adic characters of $\mathrm{Gal}(\overline{\mathbb{F}}_q/\mathbb{F}_q)$, where σ_j' and σ_j'' are irreducible ℓ-adic representations of dimension $\leq r_j$ of $\pi_1(X, \overline{\eta})$ with determinants of finite order, and where the m_{λ}'s are integers which are all zero except for finitely many λ's.

Therefore the induction hypothesis and the spectral expression for the average Lefschetz number

$$\frac{1}{\delta(\infty, o)} \sum_{\xi \in \infty \times o} \mathrm{Lef}(\mathrm{Frob}_{\xi}^{n}, \mathrm{Sht}_{\xi}^{r; \leq p} / a^{\mathbb{Z}}).$$

$(g = 1)$ given in Section 4 imply that:

P ROPOSITION (i) For each integer ν the kernel and the cokernel of the canonical homomorphism

$$H_{\mathrm{c}}^{\nu}(r; \leq p) \to \overline{H}^{\nu}(r; \leq p)$$

and the virtual $\pi_1(X \times X, \overline{\delta})$-module

$$[\overline{H}^{*}(r; \leq p)] - [H_{\mathrm{c}}^{*}(r; \leq p)]$$

are r-negligible.

(ii) There exists a virtual $\pi_1(X \times X, \overline{\delta})$-module $[H_{\mathrm{cusp}}^{*}(r; \leq p)]$ such that the difference

$$[H_{\mathrm{cusp}}^{*}(r; \leq p)] - \frac{1}{r!} \sum_{n=1}^{r!} [(\mathrm{Frob}_X^{n} \times \mathrm{Id}_X)^{*} H_{\mathrm{c}}^{*}(r; \leq p)]$$

is r-negligible, and such that, for each pair (∞, o) of distinct closed points in X, each closed point $\xi \in \infty \times o$ and each positive integer n we have

$$\mathrm{Tr}_{[H^*_{\mathrm{cusp}}(r;\leq p)]}(\mathrm{Frob}^n_\xi) = \sum_{\substack{\pi \in \mathcal{A}_r(K) \\ \omega_\pi(a)=1}} q^{(r-1)n\deg(\xi)} S_\infty^{\left(-n\frac{\deg(\xi)}{\deg(\infty)}\right)}(\pi) S_o^{\left(n\frac{\deg(\xi)}{\deg(o)}\right)}(\pi).$$

Λ

Taking into account the purity of the cohomology groups $\overline{H}^*(r; \leq p)$, which follows from the Weil conjecture proved by Deligne, Lafforgue deduces by L-function arguments:

COROLLARY

. — (i) *All the irreducible ℓ-adic representations of $\pi_1(X \times X, \overline{\delta})$ which occur in $[H^*_{\mathrm{cusp}}(r; \leq p)]$ occur with a positive multiplicity and are pure of weight $2r-2$. Moreover none of them is r-negligible.*

(ii) *The ℓ-adic representations $H^\nu_{\mathrm{c}}(r; \leq p)$, $\nu \neq 2r-2$, of $\pi_1(X \times X, \overline{\delta})$ and the virtual $\pi_1(X \times X, \overline{\delta})$-module*

$$[H^*_{\mathrm{cusp}}(r; \leq p)] - \frac{1}{r!}\sum_{n=1}^{r!}[(\mathrm{Frob}^n_X \times \mathrm{Id}_X)^* H^{2r-2}_{\mathrm{c}}(r; \leq p)]$$

are all r-negligible. Λ

Now it is also easy to deduce from Drinfeld's study of the horocycles on Sht^r and the induction hypothesis that:

L EMMA For every truncation parameters $p \leq q$ which are convex enough with respect to X the kernel and the cokernel of the canonical homomorphism

$$H^{2r-2}_{\mathrm{c}}(r; \leq p) \to H^{2r-2}_{\mathrm{c}}(r; \leq q)$$

are r-negligible. Λ

Therefore the direct limit

$$H^{2r-2}_{\mathrm{c}}(r) = \varinjlim_{p} H^{2r-2}_{\mathrm{c}}(r; \leq p) = H^{2r-2}_{\mathrm{c}}(\mathrm{Sht}^r_{\overline{\delta}}/a^{\mathbb{Z}}),$$

which is an infinite dimensional representation of $\pi_1(X \times X, \overline{\delta})$, has the following property:

. — *There exists a unique finite filtration*

$$F^\bullet = ((0) = F^0 \subset F^1 \subsetneq F^2 \subsetneq \cdots \subsetneq F^{2u+1} \subsetneq F^{2u} \subsetneq \cdots \subsetneq F^T = H^{2r-2}_{\mathrm{c}}(r))$$

such that:

- *- for any non negative integer u such that $2u+1 \leq T$, F^{2u+1}/F^{2u} is the sum of all the finite dimensional ℓ-adic subrepresentations of $H^{2r-2}_{\mathrm{c}}(r)/F^{2u}$ which are r-negligible,*
- *- for any non negative integer u such that $2u+2 \leq T$, F^{2u+2}/F^{2u+1} is the sum of all the finite dimensional ℓ-adic subrepresentations of $H^{2r-2}_{\mathrm{c}}(r)/F^{2u+1}$ which do not admit any r-negligible subquotient,*
- *- if p a truncation parameter which is convex enough with respect to X and if we denote by $F^\bullet(\leq p)$ the filtration on $H^{2r-2}_{\mathrm{c}}(r; \leq p)$ which is induced by $F^\bullet$ then, for any non negative integer u the embedding*

$$F^{2u+2}(\leq p)/F^{2u+1}(\leq p) \hookrightarrow F^{2u+2}/F^{2u+1}$$

is an isomorphism. Λ

We have thus proved:

P ROPOSITION The direct sum

$$H^{2r-2}_{\text{cusp}} = \bigoplus_{u \geq 0} F^{2u+2}/F^{2u+1}$$

is a finite dimensional ℓ-adic representation of $\pi_1(X \times X, \overline{\delta})$ and, for any truncation parameter which is convex enough with respect to X we have the equality of virtual $\pi_1(X \times X, \overline{\delta})$-modules

$$[H^{*}_{\text{cusp}}(r; \leq p)] = [H^{2r-2}_{\text{cusp}}].$$

Λ

8. A fixed point formula

Let us now consider again the action of the Hecke operators. They act on $H^{2r-2}_{\text{c}}(r) = H^{2r-2}_{\text{c}}(\text{Sht}^{r}_{\overline{\delta}}/a^{\mathbb{Z}})$ and they necessarily stabilize the above canonical filtration $F^{\bullet}$. Therefore they also act on the finite dimensional ℓ-adic representation H^{2r-2}_{cusp} of $\pi_1(X \times X, \overline{\delta})$.

To finish the induction on r, at least for the everywhere unramified representations, Lafforgue proves:

. — *For each $g \in \mathrm{GL}_r(\mathbb{A})$, each pair (∞, o) of distinct closed points in $X \setminus N_g$, each closed point $\xi \in \infty \times o$ and each positive integer n we have the equality of traces*

$$\mathrm{Tr}_{H^{2r-2}_{\text{cusp}}}(g \times \mathrm{Frob}^{n}_{\xi}) = \sum_{\substack{\pi \in \mathcal{A}_r(K) \\ \omega_\pi(a)=1}} \mathrm{Tr}_\pi(f_g) q^{(r-1)n\deg(\xi)} S_\infty^{\left(-n\frac{\deg(\xi)}{\deg(\infty)}\right)}(\pi) S_o^{\left(n\frac{\deg(\xi)}{\deg(o)}\right)}(\pi).$$

C OROLLARY For each $\pi \in \mathcal{A}_r(K)$ such that $\omega_\pi(a) = 1$ there exists an everywhere unramified Galois representation $\sigma(\pi) \in \mathcal{G}_r$ such that $L_x(\pi, s) = L_x(\sigma(\pi), s)$ for all but finitely many places x. Λ

Remark : In fact, we have $L_x(\pi, s) = L_x(\sigma(\pi), s)$ for all places x, and we have the equality of virtual modules over $\mathcal{H}_K \times \pi_1(X \times X, \overline{\delta})$

$$[H^{2r-2}_{\text{cusp}}] = \sum_{\substack{\pi \in \mathcal{A}_r(K) \\ \omega_\pi(a)=1}} [\pi^K \boxtimes (\sigma(\pi)^\vee \boxtimes \sigma(\pi))(1-r)].$$

where $\mathcal{H}_K = e_K * \mathcal{H} * e_K$ is the commutative algebra of K-biinvariant functions with compact support on $\mathrm{GL}_r(\mathbb{A})$ and $\pi^K = \pi * e_K$ is the one dimensional $\mathcal{H}_K$-module associated with the everywhere unramified $\mathcal{H}$-module π. Λ

The proof of the proposition is based on the following variant of a conjecture of Deligne on the Grothendieck-Lefschetz trace formula, in the way it has been formulated and proved by Pink ([8]).

Let $\kappa \supset \mathbb{F}_q$ be a finite field and k be an algebraic closure of κ. We simply denote

$$H^*(S) = H^*(k \otimes_\kappa S, \overline{\mathbb{Q}}_\ell)$$

the ℓ-adic cohomology of a separated scheme of finite type S over κ. The Galois group $\mathrm{Gal}(k/\kappa)$ acts on $H^*(S)$ and we denote by Frob_κ the endomorphism of $H^*(S)$ which is induced by the geometric Frobenius element in $\mathrm{Gal}(k/\kappa)$.

Let S be a proper and smooth scheme of pure dimension d over κ and $U \subset S$ be a dense open subset. We denote by $\mathrm{Frob}_S : S \to S$ the Frobenius endomorphism with respect to κ, and by $\mathrm{Frob}_U : U \to U$ its restriction to U

Let $c_U = (c_{U,1}, c_{U,2}) : V \to U \times_\kappa U$ be a finite morphism. We assume that $c_{U,1} : V \to U$ is etale, so that V is also smooth of pure relative dimension d over κ. For any positive integer n we consider the Lefschetz number

$$\mathrm{Lef}(c_U \times \mathrm{Frob}_U^n) = |\{t \in V \mid c_1(t) = \mathrm{Frob}_U^n(c_2(t))\}|.$$

If $U = S$ this Lefschetz number admits a well known cohomological interpretation. The generalized codimension d cycle $c_U : V \to U \times_\kappa U$ has a cohomology class $[c_U] \in H^{2d}(U \times_\kappa U)(d)$. Moreover, by Poincaré duality any class z in

$$H^{2d}(U \times_\kappa U)(d) = \bigoplus_{i=0}^{2d} H^i(U) \otimes_{\overline{\mathbb{Q}}_\ell} H^{2d-i}(U)(d) = \bigoplus_{i=0}^{2d} H^i(U) \otimes_{\overline{\mathbb{Q}}_\ell} (H^i(U))^\vee$$

may be viewed as an endomorphism of $H^*(U)$.

. — *If $U = S$ then, for any positive integer n we have the Lefschetz trace formula*

$$\mathrm{Lef}(c_U \times \mathrm{Frob}_U^n) = \mathrm{Tr}_{H^*(S)}([c_U] \times \mathrm{Frob}_\kappa^n).$$

Λ

If $U \subsetneq S$ we may extend c_U to a finite morphism $c = (c_1, c_2) : T \to S \times_\kappa S$ by normalizing $S \times_\kappa S$ in V. The κ-scheme T is normal, proper and of pure dimension d. The morphism $c_1 : T \to S$ is generically finite and proper, but it is not necessarily finite.

We view c as a geometric correspondence on S. Its fixed point set is the closed subset

$$\mathrm{Fix}(c) = \{t \in T \mid c_1(t) = c_2(t)\}$$

of T. More generally, for each non negative integer n we set

$$\mathrm{Fix}(c \times \mathrm{Frob}_S^n) = \{t \in T \mid c_1(t) = \mathrm{Frob}_S^n(c_2(t))\} \subset T.$$

D EFINITION We say that the correspondence c stabilizes $U \subset S$ in a neighborhood of its fixed points if there exists an open subset $W \subset T$ containing $\bigcup_{n\geq 0} \mathrm{Fix}(c \times \mathrm{Frob}_S^n)$ such that

$$c_2(c_1^{-1}(U) \cap W) \subset U.$$

. — *Let us assume that U is the complementary open subset in S of a divisor with normal crossings which is the union of a finite family $(S_i)_{i\in\Delta}$ of smooth divisors. For each $I \subset \Delta$ let us set*

$$S_I = \bigcap_{i\in I} S_i.$$

(By hypothesis S_I is proper and smooth of dimension $d - |I|$ over κ.)

Then, if c stabilizes $U \subset S$ in a neighborhood of its fixed points there exist a positive integer n_0 and cohomology classes

$$z_I \in H^{2(d-|I|)}(S_I \times_\kappa S_I)(d - |I|), \ \forall \emptyset \neq I \subset \Delta,$$

such that, for any integer $n \geq n_0$ we have the Lefschetz trace formula

$$\mathrm{Lef}(c_U \times \mathrm{Frob}_U^n) = \mathrm{Tr}_{H^\bullet(S)}([c] \times \mathrm{Frob}_\kappa^n) + \sum_{\substack{I \subset \Delta \\ I \neq \emptyset}} (-1)^{|I|} \mathrm{Tr}_{H^\bullet(S_I)}(z_I \times \mathrm{Frob}_\kappa^n).$$

Moreover, if (S, U, c) varies in an algebraic family which satisfies the obvious relative variant of the above hypotheses, the integer n_0 and the cohomology classes z_I can be chosen in a uniform way. Λ

The hypotheses of the theorem are satisfied by the Hecke correspondences. More precisely let us fix $g \in \mathrm{GL}_r(\mathbb{A})$ and let ξ be a "general enough" closed point in $X \times X$. Let us take

$$S = \overline{\mathrm{Sht}}_\xi^{r; \leq p} / a^{\mathbb{Z}} \supset U = \mathrm{Sht}_\xi^{r; \leq p} / a^{\mathbb{Z}}$$

and

$$V = c_\xi^{-1}(\mathrm{Sht}_\xi^{r; \leq p} / a^{\mathbb{Z}} \times_{\kappa(\xi)} \mathrm{Sht}_\xi^{r; \leq p} / a^{\mathbb{Z}}) \subset \mathrm{Sht}_\xi^r(g)/a^{\mathbb{Z}}.$$

T HEOREM (Drinfeld [3] for $r = 2$; Lafforgue [6] for $r \geq 3$). — *The correspondence $T \to S \times_\kappa S$ which is obtained by normalizing $S \times_\kappa S$ in V stabilizes $U \subset S$ in a neighborhood of its fixed points.* Λ

References

1. V.G. D RINFELD, Varieties of modules of F-sheaves, *Functional Analysis and its Applications* **21**, (1987), 107-122.

2. V.G. D RINFELD, The proof of Petersson's conjecture for GL(2) over a global field of characteristic p, *Functional Analysis and its Applications* **22**, (1988), 28-43.

3. V.G. D RINFELD, Cohomology of compactified manifolds of modules of F-sheaves of rank 2, *Journal of Soviet Mathematics* **46**, (1989), 1789-1821.

4. L. L AFFORGUE, Chtoucas de Drinfeld et conjecture de Ramanujan-Petersson, *Astérisque* **243**, (1997), 1-329.

5. L. L AFFORGUE, Une compactification des champs classifiant les chtoucas de Drinfeld, *Journal of the AMS* **11**, (1998), 1001-1036.

6. L. L AFFORGUE, La correspondance de Langlands sur les corps de fonctions (version préliminaire), manuscript, 1999.

7. G. L AUMON, La correspondance de Langlands sur les corps de fonctions [d'après Laurent Lafforgue], Séminaire Bourbaki, 52ème année, 1999-2000, n° 873, Mars 2000.

8. R. P INK, On the calculation of local terms in the Lefschetz-Verdier trace formula and its application to a conjecture of Deligne, *Annals of Mathematics* **135**, (1992), 483-525.

CNRS AND UNIVERSITÉ DE PARIS-SUD, UMR 8628, MATHÉMATIQUE, BÂT. 425, F-91405 ORSAY CEDEX, FRANCE

Made in United States
Orlando, FL
04 November 2026

90206105R00077